Current Topics in Microbiology and Immunology

Volume 361

Current Topics in Microbiology and Immunology

Previously published volumes

Further volumes can be found at www.springer.com

Vol. 328: **van Etten, (Ed.)**:
Lesser Known Large dsDNA Viruses. 2008.
ISBN 978-3-540-68617-0

Vol. 329: **Griffin, Diane E.**;
Oldstone, Michael B. A. (Eds.): Measles. 2009
ISBN 978-3-540-70522-2

Vol. 330: **Griffin, Diane E.**;
Oldstone, Michael B. A. (Eds.):
Measles. 2009.
ISBN 978-3-540-70616-8

Vol. 331: **Villiers, E. M. de (Ed.)**:
TT Viruses. 2009.
ISBN 978-3-540-70917-8

Vol. 332: **Karasev A. (Ed.)**:
Plant produced Microbial Vaccines. 2009.
ISBN 978-3-540-70857-5

Vol. 333: **Compans, Richard W.**;
Orenstein, Walter A. (Eds.):
Vaccines for Pandemic Influenza. 2009.
ISBN 978-3-540-92164-6

Vol. 334: **McGavern, Dorian**; **Dustin, Micheal (Eds.)**:
Visualizing Immunity. 2009.
ISBN 978-3-540-93862-0

Vol. 335: **Levine, Beth**; **Yoshimori, Tamotsu**;
Deretic, Vojo (Eds.):
Autophagy in Infection and Immunity. 2009.
ISBN 978-3-642-00301-1

Vol. 336: **Kielian, Tammy (Ed.)**:
Toll-like Receptors: Roles in Infection and Neuropathology. 2009.
ISBN 978-3-642-00548-0

Vol. 337: **Sasakawa, Chihiro (Ed.)**:
Molecular Mechanisms of Bacterial Infection via the Gut. 2009.
ISBN 978-3-642-01845-9

Vol. 338: **Rothman, Alan L. (Ed.)**:
Dengue Virus. 2009.
ISBN 978-3-642-02214-2

Vol. 339: **Spearman, Paul**; **Freed, Eric O. (Eds.)**:
HIV Interactions with Host Cell Proteins. 2009.
ISBN 978-3-642-02174-9

Vol. 340: **Saito, Takashi**; **Batista, Facundo D. (Eds.)**:
Immunological Synapse. 2010.
ISBN 978-3-642-03857-0

Vol. 341: **Bruserud, Øystein (Ed.)**:
The Chemokine System in Clinical and Experimental Hematology. 2010.
ISBN 978-3-642-12638-3

Vol. 342: **Arvin, Ann M. (Ed.)**:
Varicella-zoster Virus. 2010.
ISBN 978-3-642-12727-4

Vol. 343: **Johnson, John E. (Ed.)**:
Cell Entry by Non-Enveloped Viruses. 2010.
ISBN 978-3-642-13331-2

Vol. 344: **Dranoff, Glenn (Ed.)**:
Cancer Immunology and Immunotherapy. 2011.
ISBN 978-3-642-14135-5

Vol. 345: **Simon, M. Celeste (Ed.)**:
Diverse Effects of Hypoxia on Tumor Progression. 2010.
ISBN 978-3-642-13328-2

Vol. 346: **Christian Rommel**; **Bart Vanhaesebroeck**;
Peter K. Vogt (Ed.):
Phosphoinositide 3-kinase in Health and Disease. 2010.
ISBN 978-3-642-13662-7

Vol. 347: **Christian Rommel**; **Bart Vanhaesebroeck**;
Peter K. Vogt (Ed.):
Phosphoinositide 3-kinase in Health and Disease. 2010.
ISBN 978-3-642-14815-6

Vol. 348: **Lyubomir Vassilev**; **David Fry (Eds.)**:
Small-Molecule Inhibitors of Protein-Protein Interactions. 2011.
ISBN 978-3-642-17082-9

Vol. 349: **Michael Kartin (Eds.)**:
NF-kB in Health and Disease. 2011.
ISBN 978-3-642-16017-2

Vol. 350: **Rafi Ahmed**; **Tasuku Honjo (Eds.)**:
Negative Co-receptors and Ligands. 2011.
ISBN 978-3-642-19545-7

Vol. 351: **Marcel, B. M. Teunissen (Ed.)**:
Intradermal Immunization. 2011.
ISBN 978-3-642-23689-1

Vol. 352: **Rudolf Valenta**; **Robert L. Coffman (Eds.)**
Vaccines against Allergies. 2011.
ISBN 978-3-642-20054-0

Vol. 353: **Charles E. Samuel (Ed.)**:
Adenosine Deaminases Acting on RNA (ADARs) and A-to-I Editing. 2011.
ISBN 978-3-642-22800-1

Vol. 354: **Pamela A. Kozlowski (Ed.)**:
Mucosal Vaccines. 2012.
ISBN 978-3-642-23692-1

Vol. 356: **Cornelis Murre (Ed.)**:
Epigenetic Regulation of Lymphocyte Development. 2012.
ISBN 978-3-642-24102-4

Vol. 357: **Nicholas J. Mantis (Ed.)**:
Ricin and Shiga Toxins. 2012.
ISBN 978-3-642-27469-5

Vol. 359: **Benhur Lee, Paul Rota (Eds.)**:
Henipavirus. 2012.
ISBN 978-3-642-29818-9

Klaus Aktories · Joachim H. C. Orth
Ben Adler
Editors

Pasteurella multocida

Molecular Biology, Toxins and Infection

Responsible series editor: Klaus Aktories

Springer

Editors
Klaus Aktories
Inst. für Experimentelle und Klinische
Pharmakologie und Toxikologie
Universität Freiburg
Freiburg
Germany

Ben Adler
Department of Microbiology
Monash University
Melbourne, VIC
Australia

Joachim H. C. Orth
Inst. für Experimentelle und Klinische
Pharmakologie und Toxikologie
Universität Freiburg
Freiburg
Germany

ISSN 0070-217X
ISBN 978-3-642-31016-4 ISBN 978-3-642-31017-1 (eBook)
DOI 10.1007/978-3-642-31017-1
Springer Heidelberg New York Dordrecht London

Library of Congress Control Number: 2012943375

Printed on acid-free paper

Springer is part of Springer Science+Business Media (www.springer.com)

Preface

It is more than 125 years since Louis Pasteur identified as the causative agent of "choléra des poules" the bacterium that now bears his name. Since that time, *Pasteurella multocida* has been recognized as the causative agent of a wide range of infections in a wide range of animal species as well as in humans. Indeed, the species name *multocida* could be literally translated as "multi-killer". The term pasteurellosis is now widely applied to describe infections in mammals and birds caused by *P. multocida*. It is therefore surprising that a detailed knowledge of the cellular and molecular basis for pathogenesis in pasteurellosis lags behind what has been elucidated for many other bacterial species. In particular, the molecular basis for host predilection remains completely unknown. For example, why does a B:2 strain kill cattle, buffaloes (or mice) at a very low dose, yet seems to have no effect when injected into chickens, even at very high doses? One of the reasons for this dearth of information may well be that the construction of defined targeted mutants, the standard trade tool of molecular bacteriology, has only become routinely available in the last decade or so. Nevertheless, the development of such tools and the availability of genome sequences in recent years have facilitated significant advances in the understanding of the biology and pathogenesis of this enigmatic pathogen. A significant exception to this lack of knowledge is the mode of action of *Pasteurella multocida* toxin (PMT), a key virulence factor specifically in atrophic rhinitis. The mechanisms involved in the interaction of PMT with host cells and target molecules are now understood in exquisite detail.

This volume brings together contributions from experts in the field of *Pasteurella* research. It covers areas such as comparative genomics, pathogenic mechanisms, bacterial proteomics, as well as a detailed description and analysis of PMT and its interaction with host tissues, cells, immune system, and signaling pathways. Each review article is intended to provide a stand-alone treatise. There is, therefore, some unavoidable overlap between the various chapters.

19 March 2012

Klaus Aktories
Joachim H. C. Orth
Ben Adler

Contents

Contributors

B. Adler Department of Microbiology, Australian Research Council Centre of Excellence in Structural and Functional Microbial Genomics, Monash University, Clayton, VIC 3800, Australia, e-mail: ben.adler@monash.edu

Klaus Aktories Institut für Experimentelle und Klinische Pharmakologie und Toxikologie, Albert-Ludwigs-Universität Freiburg, Albertstr. 25, 79104 Freiburg, Germany, e-mail: klaus.aktories@pharmakol.uni-freiburg.de

J. D. Boyce Department of Microbiology, Australian Research Council Centre of Excellence in Structural and Functional Microbial Genomics, Monash University, Clayton, VIC 3800, Australia

M. Harper Department of Microbiology, Australian Research Council Centre of Excellence in Structural and Functional Microbial Genomics, Monash University, Clayton, VIC 3800, Australia

Marina Harper Department of Microbiology, Australian Research Council Centre of Excellence in Structural and Functional Microbial Genomics, Monash University, Clayton, VIC 3800, Australia

Mengfei Ho Department of Microbiology and the Host-Microbe Systems Theme of the Institute for Genomic Biology, University of Illinois at Urbana-Champaign, Urbana, IL 61801, USA

Yasuhiko Horiguchi Department of Molecular Bacteriology, Research Institute for Microbial Diseases, Osaka University, Yamada-oka 3-1, Suita, Osaka 565-0871, Japan, e-mail: horiguti@biken.osaka-u.ac.jp

Katharina F. Kubatzky Department of Infectious Diseases, Medical Microbiology and Hygiene, University Hospital Heidelberg, Im Neuenheimer Feld 324, 69120 Heidelberg, Germany, e-mail: kubatzky@uni-heidelberg.de

Alistair Lax Department of Microbiology, King's College London Dental Institute, London, UK, e-mail: alistair.lax@kcl.ac.uk

Joachim H. C. Orth Institut für Experimentelle und Klinische Pharmakologie und Toxikologie, Albert-Ludwigs-Universität Freiburg, Albertstr. 25, 79104 Freiburg, Germany, e-mail: joachim.orth@pharmakol.uni-freiburg.de

T. Seemann Department of Microbiology, Victorian Bioinformatics Consortium, Monash University, Clayton, VIC 3800, Australia

Brenda A. Wilson Department of Microbiology and the Host-Microbe Systems Theme of the Institute for Genomic Biology, University of Illinois at Urbana-Champaign, Urbana, IL 61801, USA, e-mail: bawilson@life.illinois.edu

I. W. Wilkie Department of Microbiology, Australian Research Council Centre of Excellence in Structural and Functional Microbial Genomics, Monash University, Clayton, VIC 3800, Australia

Pasteurella multocida: Diseases and Pathogenesis

I. W. Wilkie, M. Harper, J. D. Boyce and B. Adler

Abstract *Pasteurella multocida* is an enigmatic pathogen. It is remarkable both for the number and range of specific disease syndromes with which it is associated, and the wide range of host species affected. The pathogenic mechanisms involved in causing the different syndromes are, for the most part, poorly understood or completely unknown. The biochemical and serological properties of some organisms responsible for quite different syndromes appear to be similar. Thus, the molecular basis for host predilection remains unknown. The recent development of genetic manipulation systems together with the availability of multiple genome sequences should help to explain the association of particular pathological conditions with particular hosts as well as helping to elucidate pathogenic mechanisms.

Contents

I. W. Wilkie · M. Harper · J. D. Boyce · B. Adler (✉)
Department of Microbiology,
Australian Research Council Centre of Excellence in Structural and Functional Microbial Genomics, Monash University,
Clayton, VIC 3800, Australia

Current Topics in Microbiology and Immunology (2012) 361: 1–22
DOI: 10.1007/82_2012_216

Published Online: 25 May 2012

1 Non-Specific Disease Syndromes Caused by *Pasteurella multocida*

In humans, opportunist infections of soft tissue by *Pasteurella multocida* are relatively common, particularly in the elderly and immunocompromised (Baillot et al. 2011). It is a common commensal in the oropharyngeal microflora of cats and dogs, and is frequently isolated from cat bite abscesses in cats and humans (Freshwater 2008). It was reported in one study that in 50% of dog bites and 75% of cat bites the wound was contaminated with *P. multocida* (Talan et al. 1999). *P. multocida* is capable of inducing acute necrosis of adjacent tissues, and has the capacity to spread rapidly along fascial planes (Chang et al. 2007; Glass and Panigrahy 1990). The pathogenesis of wound infection by *P. multocida* is only superficially understood. In chickens and turkeys, strains which do not readily cause systemic disease (see below) are able to grow rapidly if introduced into subcutaneous tissue, causing small granulomas in chickens and sloughs in turkeys (Derieux 1978). Some known and conjectured mechanisms of tissue infection are discussed below.

2 Specific Disease Syndromes

Pasteurella multocida is associated with several distinct disease syndromes. These can be classified into two general groups:

(a) Diseases in which *P. multocida* is the primary cause, or has the principal role in the disease process. These include avian cholera, haemorrhagic septicaemia in ungulates, atrophic rhinitis in pigs and snuffles in rabbits.
(b) Diseases in which *P. multocida* makes a major contribution, although it requires other, poorly understood factors for the condition to develop. This group includes lower respiratory tract infections of ungulates and sporadic septicaemias that involve different capsular types, including non-haemorrhagic septicaemia causing type B.

Disease syndromes caused by *P. multocida* are summarised in Table 1 and each is described in detail below.

Table 1 Disease syndromes caused by *Pasteurella multocida*

Disease	Capsular serogroup	Species affected	Virulence factors
Animal bite-associated sepsis	Usually not determined, but where given usually A, occasionally D	Any, but most commonly humans, cats, dogs	Not established
Haemorrhagic septicaemia	B, E	Cattle, buffalo, pigs, goats, deer, camels, rarely horses	Capsule, fimbriae, filamentous haemagglutinin
Avian cholera	A, F, very rarely D	Chickens, turkeys, ducks, wild birds	Capsule, LPS, iron acquisition proteins, fimbriae, filamentous haemagglutinin, sialic acid uptake
Atrophic rhinitis	D, rarely A	Pigs, rabbits	*Pasteurella multocida* toxin (PMT)
Snuffles	A, D	Rabbits	Not established
Enzootic pneumonia, shipping fever	A, D, rarely F	Cattle, sheep, pigs	Not established. Co-infection with other bacterial or viral species important

3 Avian Cholera

Avian cholera, or fowl cholera (FC), is a systemic disease to which all bird species appear to be susceptible, although some species such as turkeys are particularly vulnerable. It is caused largely by strains of capsular type A, although other capsular types are occasionally incriminated, particularly type F in turkeys.

The epidemiology of avian cholera outbreaks has been investigated and the most likely source of the bacterium is other birds (Mbuthia et al. 2008). The carriage rate in flocks not experiencing outbreaks appears to be extremely low (Aye et al. 2001; Curtis and Ollerhead 1981; Muhairwa et al. 2000) and the factors which promote shedding of infective doses of bacteria by carriers are unknown. *P. multocida* survives for relatively short periods when seeded into mud or pond water, yet may be readily isolated from pond water during epornitics in waterfowl (Backstrand and Botzler 1986). Spread between birds during outbreaks is facilitated by sharing water troughs or ponds (Pabs-Garnon and Soltys 1971). Birds of prey have been infected by close contact with, or consumption of, infected water birds (Hindman et al. 1997; Williams et al. 1987).

The route of infection is not known for certain, but circumstantial evidence points toward the respiratory tract as the main site of entry. Virulent bacteria appear to be capable of colonising the mucosa of the upper respiratory tract (URT), from where they spread to the air sacs and lungs. By an unknown mechanism, but possibly via trafficking through mucosal macrophages, *P. multocida* is also able to enter the circulation directly from the mucosa (Hassanin et al. 1995; Matsumoto et al. 1991). Other possibilities for entry into the host exist. An outbreak of FC in ducklings was associated with abrasions on the legs of affected birds. *P. multocida* was readily cultured from oedematous swellings around the abrasions on birds not yet showing signs of systemic disease, and none of the dead ducklings had any lesions in their respiratory tracts, suggesting the main portal of entry in this case was via the leg abrasions (our unpublished observations).

There is wide variation in the virulence of different isolates of *P. multocida* (Petersen et al. 2001; Wilkie et al. 2000) and while all birds are likely to be susceptible to the more virulent strains, there is also variation in the susceptibility of different bird species to a given isolate of *P. multocida*. Turkeys are considered the most susceptible of the domesticated poultry species, and chickens usually the most resistant (Saif 2008). *P. multocida* may establish in the lungs by direct spread from the URT, and/or multiply in tissues, particularly liver and spleen (Rhoades and Rimler 1990). Virulent organisms experimentally introduced intravenously are initially cleared from the blood, but reappear in the later stages of the disease (Tsuji and Matsumoto 1990). Terminal-stage bacteraemias vary from moderate to high at the time of death, depending on the strain of *P. multocida* involved (Boyce et al. 2002). Bacteraemias tend to be very high with North American strains such as X-73, but can be two orders of magnitude lower with highly virulent Southeast Asian strains (our unpublished observations). The amount of visible tissue damage in acute cases at death is quite variable, which may be due to the mechanisms involved. FC strains of *P. multocida* do not produce any known toxins, and it has been suggested that, as in bovine respiratory complex disease, much of the tissue damage may be due to the host's own granulocytes (Bojesen et al. 2004). Since it requires some time to mount an effective leukocyte response, the degree of tissue damage will be dependant on the infecting dose, on how rapidly disease progresses and on the individual host response.

Lung lesions occur consistently after experimental intratracheal challenge, and are also common in field cases, a finding consistent with the respiratory tract being the principal route of infection in the natural disease. Lung lesions typically consist of substantial exudation of fluid and fibrin with necrosis and inflammation of lung tissue, and death is presumably due to respiratory failure in these birds. However, in some field cases, and in experimental cases where the organism was introduced parenterally, there is usually no evident pulmonary involvement and lesions are confined to other viscera, particularly the liver. These lesions consist of multi-focal necrosis associated with colonies of bacteria and inflammatory infiltrates. In per-acute cases, gross lesions may be restricted to small haemorrhages in the subcutis and on viscera. Multifocal haemorrhages on the epicardium with or without a fibrinous exudate and haemorrhages in coronary fat occur commonly (Fig. 1). Haemorrhages on the intestinal peritoneum and the surfaces of other viscera are also common (Saif 2008).

Another relatively common lesion of avian cholera is a mucoid exudate in intestines. This lesion seems to be common in infection with North American strains (Hunter and Wobeser 1980), but is rare with Southeast Asian strains (our unpublished observation). Some birds may develop exudative arthritis without

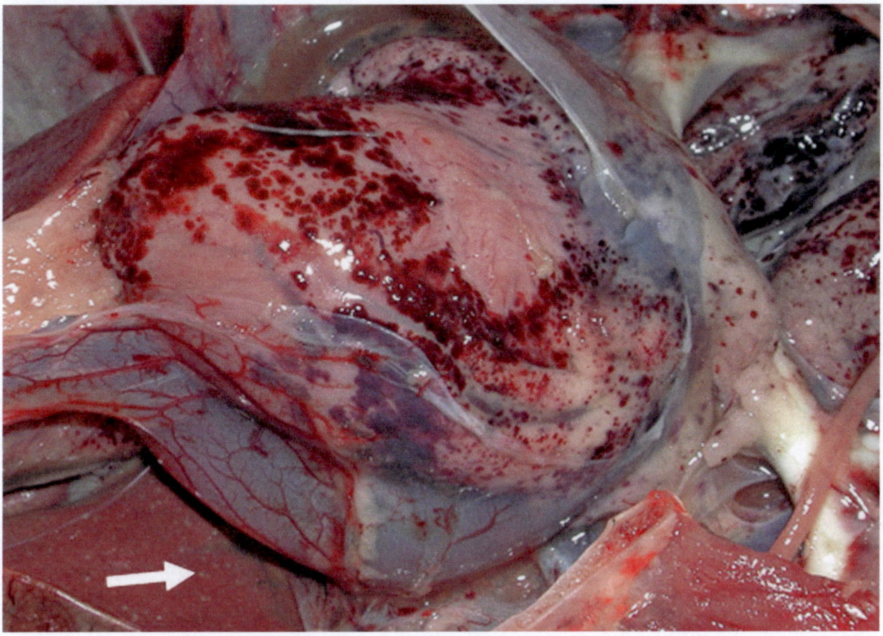

Fig. 1 Field case of fowl cholera in a duck. Note multiple haemorrhages on serosal surfaces, particularly on the epicardium. There are also numerous small white foci of necrosis in the liver (*arrow*)

showing signs of systemic disease, while inflammation of wattles and combs, and sinusitis, are also common lesions with subacute to chronic *P. multocida* infections (Saif 2008).

The clinical course of avian cholera ranges from a few hours to several days. In peracute cases, unless the birds are closely observed, the first sign may be sudden death. In many birds, the disease runs a more protracted course, ranging from around 15 h, to several days. Clinical signs include depression, lethargy, inappetence, fever and serous exudation from nostrils. Swelling of wattles and combs is commonly reported in subacute cases. In cases with significant pulmonary involvement, there will be loud respiratory rales and coughing as the disease progresses. Depending on the particular strain of *P. multocida* involved, there may be high to very high morbidity and mortality. With less virulent strains, some affected bird may slowly recover, after a variable period of depression. With more virulent strains, death usually occurs swiftly after a brief period of prostration, accompanied by convulsive wing flapping and paddling. Birds which survive the acute disease may recover completely, or may develop an exudative arthritis in leg or wing joints. Arthritis may occur without signs of acute systemic illness, particularly in very young or old birds.

4 Haemorrhagic Septicaemia

Haemorrhagic Septicaemia (HS) is a systemic disease to which all ungulates are probably susceptible, though some species (e.g. *Bubalus* spp.) appear to be more susceptible than other domesticated stock (De Alwis and Vipulasiri 1980). Livestock most commonly affected are buffalo, cattle, pigs and camels. It is caused by type B (Asia and Europe) or type E (Africa) capsular strains of *P. multocida* (De Alwis 1999). Occasionally, diseases with clinical and pathological similarities to HS have been reported, from which capsular type A strains were recovered (Cameron et al. 1996; Mackie et al. 1992).

The strains of *P. multocida* which cause HS have been isolated from a wide range of animals, both domesticated and wild, but do not appear to be capable of extended survival outside the host (De Alwis 1999). It is well-recognised in countries where HS is endemic that small numbers of clinically normal animals carry virulent strains of HS-causing *P. multocida*. These strains can be isolated from tonsils of healthy animals, even those with high serum titres for *P. multocida* antigens (De Alwis et al. 1986; Sheikh et al. 1994). What causes carriers to become shedders is not known. A carrier animal may shed intermittently and repeatedly without contracting disease itself, or may develop fulminating HS after previous episodes of shedding without signs of disease. Outbreaks occur when carriers or diseased animals shed large numbers of organisms into the immediate environment of susceptible animals through nasal and oral excretions (De Alwis 1999).

In the natural disease, the main focus at which the causative organism enters is considered to be the tonsils, via macrophages and their so-called "tonsilar sampling." Bacteria multiply in peritonsilar soft tissue, and macrophages may also carry viable organisms to local lymph nodes, where they multiply and spread to surrounding tissues. Once established, the bacteria grow rapidly, causing marked local oedema and tissue necrosis.

Like all *P. multocida*, HS-causing strains are capable of rapid growth once established in soft tissue. The affected animal mounts an inflammatory response to the infection, but it is incapable of arresting bacterial growth. The combination of tissue necrosis, local and systemic cytokine induction, and developing endotoxaemia, induces the clinical signs of a rapidly fulminating acute disease. A bacteraemia develops, but there is no convincing evidence that the bacteria actually multiply in the blood in vivo, so the term "septicaemia" for the disease may be a misnomer. Some investigators claim that the entire clinical presentation is due to effects of endotoxin alone (Horadagoda et al. 2001). While clinical signs may be replicated with endotoxin, these manifest in the latter stages of disease, so it is unlikely that endotoxin contributes significantly to the initial pathogenesis. The details of how and why bacteria gain access to replication sites and how they manage to rapidly multiply in what should be a hostile environment remain to be adequately explained.

Because first-hand experience of the field disease has not been well-documented, most descriptions of HS are of experimental disease. These have been described in a number of publications, but perhaps the most reliable is given in the monograph by De Alwis (1999). In peracute cases the first sign may be sudden death, but animals kept under close observation will exhibit fever, inappetance, nasal discharge, excessive salivation and laboured breathing. These signs are usually accompanied by swelling in the sub-mandibular region, which may extend to the brisket and forelegs. As the disease progresses, respiratory difficulty increases, nasal discharge changes from mucoid to mucopurulent. Eventually, the affected animal becomes recumbent and dies. The mortality rate without prompt antibiotic treatment is close to 100%.

5 Atrophic Rhinitis

Atrophic Rhinitis (AR), a syndrome which may include facial distortion, turbinate bone destruction and retarded growth of young pigs, has long been recognised as a common condition of domesticated pigs in all countries. It also occurs with some frequency in housed rabbits (DiGiacomo et al. 1991). The syndrome has also been reported as a spontaneous disease of wild boar (Kaden et al. 2001), goats (Baalsrud 1987) and cattle (Rouman and Van de Goor 1988). An experimental model has been developed in mice (Jordan and Roe 2004). The clinical effects described are attributed to a so-called dermonecrotoxin, often referred to as PMT (for "*Pasteurella multocida* toxin"), which is associated with certain strains of *P. multocida*, mostly capsular type D, but other capsular types may also produce PMT. Details of the action of PMT are described in Molecular Biology of *Pasteurella multocida* Toxin.

PMT-producing *P. multocida* does not appear to colonise the URT readily. Multiple inoculations, or the presence of other respiratory tract commensals such as *Bordetella bronchiseptica* were necessary to establish toxigenic *P. multocida* on the nasal mucosa of gnotobiotic piglets (Eliás et al. 1990). Toxigenic type D strains do possess fimbriae, but toxin production seems to be more important for adhesion. A toxigenic type A strain was found to be equally adhesive in the absence of fimbriae (Pijoan and Trigo 1990). Alteration of the mucosal environment by chemicals such as ammonia can also assist colonisation of the mucosa by *P. multocida* (Hamilton et al. 1998). After successfully colonising the URT, toxin-producing bacteria may remain on the mucosa and tonsils for some time, thereby providing a continuing source of PMT (Ackermann et al. 1991).

The syndrome of AR occurs after the URT becomes colonised by PMT-producing *P. multocida*. The bacterium does not need to invade the host tissues to any significant extent to cause lesions. Pure PMT instilled in the nasal cavity causes osteopaenia and destruction, beginning in the ventral turbinate bones (Dominick and Rimler 1988; Martineau-Doize et al. 1991). This is initiated by modulation of osteoblasts, which in turn affect the activity of osteoclasts (Mullan and Lax 1998). The net result is loss of turbinate structure and distorted facial growth. PMT has also been associated with suppressed bodyweight gain under experimental conditions (Eliás et al. 1990). It has also been suggested that loss of turbinates predisposes pigs to more frequent and severe lower respiratory tract infection because of the reduced filtering of incoming air. However, the extent of any detrimental effects is difficult to demonstrate in farmed pigs. Morrison et al. (1985) found a small correlation between severity of turbinate damage and the extent of pneumonia, but no measurable effect on growth rates, while Paisley et al. (1993) were unable to find any correlation between severity of AR and growth rates in Danish pigs. In a comparison of three herds in the United States, one herd did show an association between severity of turbinate lesions and growth rate, but in a combined analysis of the three herds, the correlation was not maintained (Scheidt et al. 1990).

The main clinical signs of AR include distorted maxillae, deviation of nasal septum, erosion or absence of nasal turbinates, and possibly, depressed growth rate, though the latter effect may be associated with management practices more than the disease. Infections acquired after growth is complete are unlikely to cause any facial deformities. The lesions of bony tissue are permanent, although the mucosa may repair. Most pigs are slaughtered at a relatively young age, so it is not known what proportion, if any, eventually eliminate the toxigenic strains from their oropharyngeal tissues.

6 Snuffles of Rabbits

Snuffles is an acute to chronic exudative rhinitis, which, particularly in neglected cases, may lead to a fulminating, suppurative pneumonia. It is associated with colonisation of the URT with *P. multocida*, mostly capsular type A, but type D may also be present, either in combination with type A or alone. The most common route of infection is probably direct, from animal to animal under the close contact conditions of housed rabbits (DiGiacomo et al. 1991). As with atrophic rhinitis, *B. brochiseptica* may play an important role in establishment of *P. multocida* and severity of lesions (Deeb et al. 1990). Interestingly, atrophic rhinitis has been reported in rabbits in association with capsular type A in the absence of type D (DiGiacomo et al. 1989).

The respiratory mucosa responds in accordance with the degree of damage, and other bacteria such as *B. bronchiseptica* contribute to establishment of *P. multocida* and tissue damage (Deeb et al. 1990). Some infective material is inevitably aspirated, and *P. multocida* may establish in the lower respiratory tract, causing either a chronic, localised pneumonia or a fulminating and fatal suppurative pneumonia (Dillehay et al. 1991). The first clinical sign of rhinitis is a serous nasal discharge, which progresses to a mucopurulent discharge, as the severity of rhinitis increases. The rhinitis may eventually regress. In some rabbits, bacteria spread to the lungs where they may cause a localised, chronic lobar pneumonia or a severe, fulminating, fibrinopurulent pneumonia and pleuritis, in which case, clinical signs will include coughing, fever, mild or severe respiratory difficulty, and eventually death.

7 Lower Respiratory Tract Infections of Ungulates

Lower respiratory tract (LRT) infections (pneumonia and pleuritis) associated with *P. multocida* may manifest as acute, fulminating conditions (e.g. so called "shipping fever" of cattle) or chronic, lobar pneumonias such as the so-called "enzootic pneumonias" in housed swine, calves and sheep. *P. multocida* is not the sole cause, and not the only bacterial species associated with these types of pneumonias, but it is one of the most commonly isolated species from pneumonias of swine (Hansen et al. 2010). While not as frequently involved as *Mannheimia haemolytica* in acute fibrinous pneumonias of cattle, *P. multocida* is a very common contributor to the condition (Dabo et al. 2008) and appears to be increasingly involved in acute fibrinous pneumonias of cattle in Britain (Veterinary Investigation Diagnosis Analysis, and Veterinary Laboratories Agency, UK, statistics cited in Dowling et al. (2002)). The most commonly isolated strains are of capsular type A, but other capsular types are sometimes isolated from pneumonic lungs, such as types D and occasionally F (Bethe et al. 2009).

Despite much research effort over the last 80 or more years, the pathogenesis of pneumonic pasteurellosis in all species remains enigmatic. Environmental factors plus prior and concurrent infections with viruses and other bacteria have been recognised as contributing to the establishment of *P. multocida* in the LRT of cattle (Fulton 2009). How and why organisms switch from being commensal URT organisms to pathogens is still not understood. Non-serogroup D *P. multocida* produces no toxin, but once established in the LRT, it provokes an extreme inflammatory response and lesions that are morphologically indistinguishable from those associated with *M. haemolytica*. In housed swine and calves, pneumonias associated with *P. multocida* may be acute and fulminating, but tend most often to be more localised, less exudative, less destructive and chronic. However, the underlying reasons for the differences are not known.

Oropharyngeal carriage of *P. multocida* is common in all of the species involved, and aspiration of small numbers of bacteria is inevitable. In swine, prior lower respiratory infection with species of *Mycoplasma* allows *P. multocida* to establish and cause lesions more readily in the lower respiratory tract (Ciprian et al. 1988). Similarly, a variety of prior infections with viruses or bacteria may allow the establishment of *P. multocida* in the LRT of cattle (Confer 2009).

There is no information on the survival of *P. multocida* in the livestock housing environment. Analysis of strains recovered from pigs supports the view that there is a slow exchange of bacteria from animal to animal within groups (Zhao et al. 1993). Several biovars are concurrently present in each group and even within individuals (Bowles et al. 2000; Harel et al. 1990).

The lung lesions caused by *P. multocida* are similar or identical to those caused by several other bacterial pathogens. There are virulence factors common to the various pathogens and some that are unique to *P. multocida*, but it is not known which are important for development of respiratory disease. An important step in disease initiation is thought to be altered clearance and killing of bacteria in the LRT. This is normally a very efficient process, and experimental induction of pneumonia in any of the domesticated species requires inoculation of very high numbers of organisms (Hall et al. 1990) or various pre-treatments to render the animals more susceptible to infection (Ross 2006). Pulmonary macrophages are considered to be primarily responsible for clearance of organisms from bronchioles and alveoli, so it is reasonable to assume that an alteration in the ability of pulmonary macrophages to phagocytose or kill bacteria is part of the pathogenesis. However, it is difficult to demonstrate changes in macrophage function in vivo and while in vitro experiments have demonstrated that pre-treatments can cause differences in macrophage function, these have not resolved the early pathogenesis of infection (Muller and Kohler 1997; Purdy et al. 2003).

Once clinical disease begins to develop in adult cattle, it is accompanied by a marked neutrophil response at the site of infection, which plays a very important role in the severity of the lung lesions. What provokes the neutrophil response to *P. multocida* remains to be completely explained. *M. haemolytica* produces a specific leukotoxin which is highly toxic to bovine neutrophils via their CD18 surface receptors (Despande et al. 2002), but *P. multocida* produces no detectable leukotoxins, yet still manages to provoke a very significant neutrophil infiltrate and lesions that are morphologically identical to those associated with *M. haemolytica*. In chronic forms of pulmonary pasteurellosis, the less severe lesions contain fewer neutrophils and a greater proportion of macrophages.

Inflammatory cell responses are regulated by cytokines, but as yet few studies have been designed that analyse the precise cytokine sequence and the effect of local and systemic cytokine responses in pneumonic pasteurellosis (Berndt et al. 2002). It is likely that dysregulation of inflammatory cytokines is a common factor in the development of these pneumonias.

In severe, acute pneumonias, the exudation and tissue necrosis rapidly increases within lobes by direct spread, and between lobes by aspiration of contaminated material forced into the bronchi. Without treatment, the affected animal becomes feverish, displays marked respiratory difficulty with mouth-breathing and frothing at the mouth due to expectorated material containing surfactant. Death from respiratory failure is likely within a few days from the clinical onset of pneumonia. The typical pathological changes include consolidation of the cranioventral lobes with evident haemorrhage and necrosis, with a heavy fibrin exudate on the pleural surface and in the interlobular septae. Small to medium airways are often blocked by inspissated necrotic material consisting mainly of disintegrating inflammatory and somatic cells (Jubb and Maxie 2007). Chronic pneumonias usually run a much more protracted course, and clinical signs may be absent or mild. The lung lesions may heal, or slowly progress to the point that they cause respiratory difficulty.

8 Sporadic Septicaemias Involving Different Capsular Types

There are numerous reports of toxaemic/septicaemic disease associated with *P. multocida*. The strains involved expressed a range of capsule types and were not typical type B or type E HS-associated strains, Outbreaks are sporadic and the species most commonly involved are pigs and calves, and less frequently, lambs and horses. Affected animals die from septic shock, and the clinicopathological features are very similar to HS. A typical example is that reported by Cameron et al. (1996) in which an outbreak of septicaemia in a large pig herd was associated with *P. multocida* subspecies *gallicida*, (capsule type A). The organism was recovered in large numbers from most tissues cultured, but no disease was caused in healthy pigs when even very large numbers of the recovered organisms were administered either intranasally or intravenously (I. Wilkie, unpublished data). An outbreak of a similar clinical and pathological condition in pigs was associated with *P. multocida* capsular type D (Mackie et al. 1992). Although some possible predisposing factors were identified, including sudden change in ambient temperature, a dietary change and high levels of manure gas, it was not determined which, if any, predisposed the pigs to the disease outbreak.

There have been reports of outbreaks of serositis and septicaemia in calves associated with a strain expressing capsular type F in Germany (Catry et al. 2005), and capsular type B in New Zealand (McFadden et al. 2011). Outbreaks in neonatal lambs due to type F (Watson and Davies 2002) and an untyped strain (Rad et al. 2011) have also been reported. The common feature of all of these conditions is the sudden onset of a fulminating toxaemic disease with rapid prostration and death from what is essentially septic shock. Crowding and other aspects of poor management such as exposure to cold, wet conditions are suspected as contributing factors, but which elements are most important in initiating an outbreak are unknown. What remains to be discovered is whether the sudden acquisition of apparent virulence of the associated *Pasteurella* species is due to upregulation of virulence factors in response to environmental factors, or whether host immunity is compromised, allowing commensal organisms to flourish unchecked. The wide range of strains responsible for these sporadic septicaemias suggests that changes in the host are probably most important in initiating the syndrome. However, the possibility that the strains involved do carry essential virulence factors, or require a particular combination of factors, should also be considered.

Oropharyngeal carriage of strains of *P. multocida* is common, so it is likely that some of the individuals within a group initially carry the organisms as part of their commensal microflora. Two reported incidents involved strains belonging to capsular group F (Catry et al. 2005; Watson and Davies 2002), which is not commonly associated with mammals, so in this case the causative organism could also have been acquired from some other species that had been in recent contact.

The pathogenesis of this syndrome is unknown. It would appear that a nidus of infection establishes at some site such as lungs, pleura or peritoneum, from where it spreads either directly or into the circulation and is carried to multiple sites.

Clinical disease develops rapidly from some point, probably when endotoxin levels reach a critical concentration. There is usually a very short course of depression, fever, prostration and death due to septic shock.

9 Mechanisms of Pathogenesis: Cellular and Molecular Aspects

Given the very broad spectrum of diseases caused by *P. multocida* and their worldwide economic impact, it may appear surprising that relatively little is known about the pathogenesis of pasteurellosis at the cellular and molecular level. This is especially apparent when comparisons are made with the detailed understanding of pathogenesis in other Gram-negative pathogens such as *Salmonella, Shigella* and *Yersinia*. The principle reason for this lack of knowledge is the fact that efficient genetic manipulation of *P. multocida* has been possible only in the last decade or so. Most of the work conducted before the genomic era was descriptive in nature; genes or proteins responsible for virulence-associated phenotypes were usually not identified or characterised. For a more detailed description of some of these earlier studies, the reader is referred to Boyce et al. (2010) and Harper et al. (2006). By far the best characterised mediator of disease is the PMT toxin carried by some strains of *P. multocida* that cause porcine atrophic rhinitis. Details of PMT are discussed in Molecular Biology of *P. multocida* Toxin.

10 Resistance to Host Innate Immunity

One of the first barriers that infecting bacteria need to overcome is the host innate immune system, comprising multiple mechanisms of defence, such as phagocytic cells, host sequestration of free iron, the bactericidal action of complement and the activity of antimicrobial peptides. The major surface components of *P. multocida*, capsule and lipopolysaccharide (LPS), play important roles in the ability of bacteria to resist these mechanisms. As detailed in The Key Surface Components of *P. multocida*: Capsule and LPS, defined acapsular mutants are attenuated for virulence because they are either readily phagocytosed and/or killed by complement (Boyce and Adler 2000; Chung et al. 2001). Likewise, truncation of the LPS molecule results in attenuation, due at least in part to increased susceptibility to the bactericidal action of avian antimicrobial peptides such as fowlicidin (Boyce et al. 2009).

Iron is essential for the growth of almost all species of bacteria. However, the concentration of free iron in mammalian and avian host tissues is too low to support bacterial growth because iron is sequestered by host proteins such as transferrin, lactoferrin and haemoglobin. Not surprisingly therefore, *P. multocida* has evolved mechanisms to obtain iron from host iron-binding proteins. Indeed, the genome of *P. multocida* encodes more than 50 proteins with predicted roles in iron acquisition and transport (May et al. 2001). Whole genome microarray studies have identified a large number of genes (up to 12%) with altered expression under iron-limiting conditions or in response to defined sole iron sources (Paustian et al. 2001, 2002). Of course, many of these genes are not involved directly in iron metabolism, but reflect physiological responses to changed environmental and nutritional conditions. Interestingly, analysis of the outer membrane sub-proteome under iron-limiting conditions identified only two proteins, OmpW and Pm803, with increased expression (Boyce and Adler 2006).

A number of *P. multocida* proteins which bind haemoglobin and/or haemin has been identified (Bosch et al. 2004). Some of them have been characterised in detail. The HgbA protein bound haemoglobin in vitro, but an *hgbA* mutant retained haemoglobin binding ability and virulence for mice (Bosch et al. 2002b). Similarly, an *hgbB* mutant bound haemoglobin to the same extent as the wild-type parent and was not attenuated in mice (Cox et al. 2003). Intriguingly, a signature-tagged mutagenesis (STM) screen identified an *hgbA* mutant which was attenuated for virulence in mice (Fuller et al. 2000). The reasons for this apparent discrepancy are unclear, but may be due to the use of different serotypes and/or different mouse strains. However, the mutation was not complemented in the STM study, so it is possible that the attenuation observed may have been due to an unrelated mutation.

The transferrin binding proteins (Tbp) of *P. multocida* have been less well characterised. Unusually for Gram-negative bacteria, *P. multocida* requires a single protein, TbpA, to bind host transferrin (Ogunnariwo and Schryvers 2001). Interestingly, in a survey of 289 isolates, the *tbpA* gene was present in only 31.5% of strains (Ewers et al. 2006) and restricted to ruminant isolates. Given that the PCR primers used in that study were based on the original TbpA from a bovine isolate, it is likely that genes differing in sequence were not detected. No homologues of *tbpA* are present in any of the currently available genome sequences (Pathogenomics of *P. multocida*). However, the upregulation of a range of iron acquisition genes during growth with transferrin as the sole iron source (Paustian et al. 2002) is consistent with the presence of as yet uncharacterised transferring binding proteins.

The lack of attenuation in almost all mutants that have been constructed in single iron acquisition genes is consistent with a high degree of redundancy for iron acquisition in *P. multocida* (Bosch et al. 2002a, b, 2004; Cox et al. 2003). Such redundancy is not uncommon. Indeed, in the related pathogen *Haemophilus influenzae* it was necessary to inactivate all three genes encoding haemoglobin-binding proteins before a phenotypic effect was observed (Morton et al. 1999). However, energy for these various iron uptake systems is derived via the common TonB protein and the accessory proteins ExbB and ExbD. Not surprisingly therefore, inactivation of any one of these genes results in attenuation of virulence (Bosch et al. 2002a; Fuller et al. 2000).

Several of these iron-acquisition proteins have been tested as potential vaccine antigens; none has elicited protection. For a more detailed description of iron acquisition in *P. multocida*, the reader is referred to Boyce et al. (2010).

11 Cell Adhesion and Invasion

It is axiomatic that adhesion to host tissues is an essential prerequisite for the establishment of infection. There have been many studies which reported adhesion of *P. multocida* to a range of cell types from mammalian and avian hosts. However, most of the early work did not identify either the bacterial component(s) involved or the corresponding host receptor(s). They will not be discussed here.

Given that the hyaluronic acid (HA) capsular polysaccharide of serogroup A *P. multocida* is indistinguishable from mammalian and avian tissue HA, it is perhaps a little surprising that *P. multocida* HA can mediate adhesion to host cells. Nevertheless, bacterial adhesion to both turkey air sac macrophages and also to turkey monocytes has been reported as being mediated by capsular HA (Pruimboom et al. 1999). The monocyte receptor was identified as CD44, a cell-surface glycoprotein with a role in intercellular interactions, that acts as a receptor for extracellular matrix (ECM) HA. The conclusion was based on inhibition of adhesion by HA or by an anti-CD44 monoclonal antibody; paired isogenic strains were not available. The suggestion was that *P. multocida* is able to adhere to host cells by mimicking host HA. These studies have not been reproduced.

Other components of the ECM have been identified as putative receptors for the adhesion of *P. multocida.* In particular, adhesion to fibronectin has been reported by a number of workers. In common with all Gram-negative bacteria, *P. multocida* has a homologue of the major outer membrane protein, OmpA, which was shown to bind fibronectin as well as heparin on the surface of bovine kidney (MDBK) cells (Dabo et al. 2003). The adhesion of *P. multocida* cells to MDBK monolayers was inhibited by anti-OmpA antibodies. The same group reported several additional proteins that interacted with fibronectin and were therefore postulated as adhesins (Dabo et al. 2005). In that study *P. multocida* bound to fibronectin, but not to laminin or collagen, and the adhesion was inhibited by anti-fibronectin antibodies. However, fibronectin is a large, complex molecule that facilitates interactions with many other components of extracellular matrix; it is not surprising that there is in vitro evidence for binding to many bacterial proteins. Indeed, subsequent studies have identified yet more fibronectin-binding proteins in *P. multocida.* A *P. multocida* homologue of the highly conserved DNA uptake protein ComE was shown to also bind fibronectin (Mullen et al. 2008a, b) via the same helix-turn-helix motif involved in DNA binding. In the absence of isogenic strains and complementation data, along with the appropriate controls, the biological significance of these interactions is difficult to interpret.

Bacterial fimbriae (pili) are well documented as key adhesins, and therefore virulence factors, in many bacterial species. This includes type 4 fimbriae, which are the most widespread fimbrial type and are found in a very diverse range of bacterial genera (Pelicic 2008). Type 4 fimbriae have also been used successfully as vaccines (Tennent and Mattick 1994). The *P. multocida* genome contains all of the genes necessary for the biogenesis of type 4 fimbriae and their presence has been reported in serogroup A, B and D strains (Ruffolo et al. 1997). One might therefore hypothesise a key role for these structures as adhesins and virulence determinants. However, no defined genetic studies have yet been conducted which might prove (or refute) such a hypothesis.

P. multocida also contains all the genes required for the synthesis and assembly of Flp pili, a sub-family of type 4 fimbriae first identified as an adhesion in the related species *Actinobacillus* (*Aggregatibacter*) *actinomycetemcomitans* (Kachlany et al. 2001). These structures have not been identified in *P. multocida*, but there is compelling evidence for their expression and role in virulence. Two independently performed screens of STM libraries identified mutants in Flp1, the pilin subunit protein or TadD, an assembly/secretion protein. Both mutants were attenuated in mice (Fuller et al. 2000; Harper et al. 2003). Significantly, the Flp1 mutant was also attenuated for virulence in the natural chicken host. The molecular basis for attenuation has not been elucidated.

The first sequence of a *P. multocida* genome identified two paralogous genes, *fhaB1* and *fhaB2*, encoding proteins with similarity to the large filamentous haemagglutinin of *Bordetella pertussis* (May et al. 2001). Given the large sizes of both genes (7.8 and 11.7 kb respectively), it was not surprising that STM mutants in both genes were obtained previously in transposon mutant libraries. Both mutants were attenuated for virulence in mice (Fuller et al. 2000). A subsequently constructed directed mutant in *fhaB2* was slightly attenuated in turkeys (Tatum et al. 2005). Interestingly, following intravenous infection there was no difference in mortality but the birds infected with the mutant showed a longer survival time. Intranasal infection resulted in an apparent increase of approximately 100-fold in the lethal dose of the mutant, but these conclusions must be tempered by the fact that only two out of six control birds died. Nevertheless, the results were consistent with FhaB2 being an adhesin that contributes to, but is not essential for, virulence.

P. multocida is not considered to be an intracellular pathogen. However, there have been several reports of invasion of a variety of cell types by *P. multocida*. For example, a virulent serotype A:3,4 strain was able to enter primary turkey kidney cells at a level tenfold higher than the attenuated A:3,4 vaccine strain (Lee et al. 1994). Likewise, a serogroup B strain was shown to be internalised by primary bovine endothelial cells (Galdiero et al. 2001), but the bacteria did not replicate intracellularly and caused no damage to the cells. The biological significance of these reports thus remains speculative. Similarly, the claimed role of capsule in the apparent invasion of chicken fibroblasts by serogroup A strains must be tempered by the use of non-isogenic strains and the fact that treatment with hyaluronidase had no effect on internalisation (Al-haj Ali et al. 2004).

While it is clear that the capsule can play a key role in the ability of *P. multocida* to resist phagocytosis, there are conflicting data on whether, once internalised, *P. multocida* is able to replicate inside macrophages or granulocytes. Following opsonisation by immune serum, virulent *P. multocida* has been shown in a number of studies to be readily killed by turkey macrophages or heterophils (Harmon et al. 1992), chicken macrophages (Poermadjaja and Frost 2000) and mouse macrophages (Collins et al. 1983). On the other hand, we have demonstrated survival, but

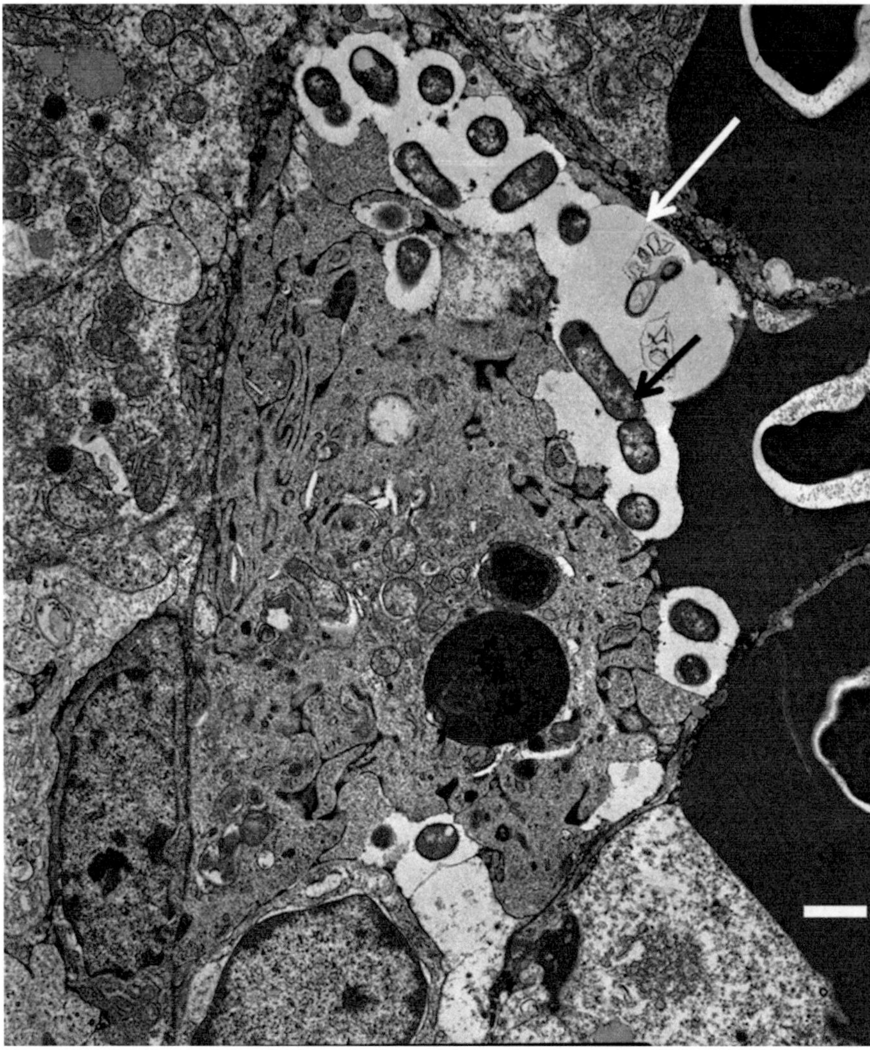

Fig. 2 Electron micrograph of fixed hepatic macrophage from a chicken 18 h after intramuscular inoculation of 10^4 *P. multocida* A:1 strain VP161. There are numerous apparently intact bacteria within membrane-bound vesicles (*white arrow*) in the cytoplasm, while the macrophage shows signs of early degeneration. One bacterium appears to be completing cell division (*black arrow*). Bar = 1 μm. Image courtesy of Majid Ghoddusi

not replication, of *P. multocida* in mouse peritoneal macrophages following phagocytosis both in the presence and absence of specific antibody (Boyce and Adler 2000; Ramdani 1992). Some of these differences may well be due to the serogroup and virulence of the strains used. In addition, it is unclear whether the results from these in vitro studies can be extrapolated to the situation in infected animals. Indeed, the fixed macrophages in the liver appear to play a pivotal role in progression to disease if bacteria enter the blood stream (Tsuji and Matsumoto 1990), suggesting that the ability of *P. multocida* to survive phagocytosis is a crucial step in the pathogenesis of systemic disease. It has been shown that *P. multocida* can multiply intracellularly after endocytosis by amoebae in a situation analogous to that in mammalian and avian macrophages (Hundt and Ruffolo 2005). Moreover, we have observed encapsulated bacteria within macrophages in liver sections of birds infected with the highly virulent A:1 strain VP161 (Fig. 2). Bacteria were found largely intact, with evidence of cell division, but macrophages appeared in varying stages of degeneration.

12 Other Virulence-Associated Factors

Two screens of STM libraries have identified a total of approximately 40 mutants which were attenuated in mice and/or chickens (Fuller et al. 2000; Harper et al. 2003). Some of these were mutants in genetic loci known previously to affect virulence, such as LPS or capsule biosynthesis, while others were metabolic mutants in genes encoding nucleotide synthesis or amino acid synthesis or transport proteins. The latter could be more appropriately considered as virulence-associated factors rather than true mediators of virulence (Wassenaar and Gaastra 2001). The facts that a number of genes of unknown function were identified, and that neither STM screen achieved genome saturation, suggest that many more virulence factors await identification.

P. multocida cannot synthesise sialic acid de novo and must therefore acquire it exogenously. The role of sialic acid metabolism in virulence has been somewhat controversial. See Boyce et al. (2010) for a more detailed review. However, the ability of *P. multocida* to take up and incorporate sialic acid is a requirement for virulence, at least in turkeys (Tatum et al. 2009). A mutant with both the *nanP* and *nanU* genes inactivated was attenuated for lethality by at least 100-fold, regardless of whether the challenge was parenteral or intranasal. The precise mechanism remains unknown, but it is intriguing to speculate that under certain conditions *P. multocida* may sialylate its LPS (see The Key Surface Components of *Pasteurella multocida*: Capsule and Lipopolysaccharide), leading to either antigenic masking or mimicry.

Acknowledgments Original work in the author's laboratories was supported by the Australian Research Council, the Australian Centre for International Agricultural Research, and the Australian Poultry Co-operative Research Centre.

References

Ackermann MR, Cheville NF, Gallager JE (1991) Colonization of the pharyngeal tonsil and respiratory tract of the gnotobiotic pig by a toxigenic strain of *Pasteurella multocida* type D. Vet Pathol 28:267

Al-haj Ali H, Sawada T, Hatakeyama H, Katayama Y, Ohtsuki N, Itoh O (2004) Invasion of chicken embryo fibroblast cells by avian *Pasteurella multocida*. Vet Microbiol 104:55–62

Aye PP, Angerick EJ, Morishita TY, Harr BS (2001) Prevalence and characteristics of *Pasteurella multocida* in commercial turkeys. Avian Dis 45:182–190

Baalsrud KJ (1987) Atrophic rhinitis in goats in Norway. Vet Rec 121:350–353

Backstrand JM, Botzler RG (1986) Survival of *Pasteurella multocida* in soil and water in an area where avian cholera is enzootic. J Wildl Dis 22:257–259

Baillot R, Voisine P, Côté L, Longtin Y (2011) Deep sternal wound infection due to *Pasteurella multocida*: the first case report and review of literature. Infection 39:575–578

Berndt A, Heller M, Kosmehl H (2002) Cytokine mRNA expression in experimental porcine pneumonia. Deutsche Tierarzt Wochenschr 109:205–209

Bethe A, Weiler LH, Selbitz H-J, Ewers C (2009) Genetic diversity of porcine *Pasteurella multocida* strains from the respiratory tracts of healthy and diseased swine. Vet Microbiol 139:97–105

Bojesen A, Petersen K, Nielsen O, Christensen J, Bisgaard M (2004) *Pasteurella multocida* infection in heterophil-depleted chickens. Avian Dis 48:463–470

Bosch M, Garrido E, Llagostera M, Perez de Rozas AM, Badiola I, Barbe J (2002a) *Pasteurella multocida exbB, exbD* and *tonB* genes are physically linked but independently transcribed. FEMS Microbiol Lett 210:201–208

Bosch M, Garrido ME, Llagostera M, Perez de Rozas AM, Badiola I, Barbe J (2002b) Characterization of the *Pasteurella multocida hgbA* gene encoding a hemoglobin-binding protein. Infect Immun 70:5955–5964

Bosch M, Garrido ME, Perez de Rozas AM, Badiola I, Barbe J, Llagostera M (2004) *Pasteurella multocida* contains multiple immunogenic haemin- and haemoglobin-binding proteins. Vet Microbiol 99:103–112

Bowles RE, Pahoff JL, Smith BN, Blackall PJ (2000) Ribotype diversity of porcine *Pasteurella multocida* from Australia. Aust Vet J 78:630–635

Boyce JD, Adler B (2000) The capsule is a virulence determinant in the pathogenesis of *Pasteurella multocida* M1404 (B:2). Infect Immun 68:3463–3468

Boyce JD, Adler B (2006) How does *Pasteurella multocida* respond to the host environment? Curr Opin Microbiol 9:117–122

Boyce JD, Wilkie I, Harper M, Paustian M, Kapur V, Adler B (2002) Genome scale analysis of *Pasteurella multocida* gene expression during growth within the natural chicken host. Infect Immun 70:6871–6879

Boyce JD, Harper M, St Michael F, John M, Aubry A, Parnas H, Logan SM, Wilkie IW, Ford M, Cox AD, Adler B (2009) Identification of novel glycosyltransferases required for assembly of the *Pasteurella multocida* A:1 lipopolysaccharide and their involvement in virulence. Infect Immun 77:1532–1542

Boyce JD, Harper M, Wilkie I, Adler B (2010) *Pasteurella*. In: Gyles CL, Prescott JF, Songer JG, Thoen CO (eds) Pathogenesis of bacterial infections in animals. Blackwell Publishing, Iowa

Cameron RDA, O'Boyle D, Frost AJ, Gordon AN, Fegan N (1996) An outbreak of haemorrhagic septicaemia associated with *Pasteurella multocida* subsp gallicida in a large pig herd. Aust Vet J 73:27–29

Catry B, Chiers K, Schwarz S, Kehrenberg C, Decostere A, de Kruif A (2005) Fatal peritonitis caused by *Pasteurella multocida* capsular Type F in calves. J Clin Microbiol 43:1480–1483

Chang K, Siu LK, Chen Y-H, Lu P-L, Chen T-C, Hsieh H-C, Lin C-L (2007) Fatal *Pasteurella multocida* septicemia and necrotizing fasciitis related with wound licked by a domestic dog. Scand J Infect Dis 39:167–170

Chung JY, Wilkie I, Boyce JD, Townsend KM, Frost AJ, Ghoddusi M, Adler B (2001) Role of capsule in the pathogenesis of fowl cholera caused by *Pasteurella multocida* serogroup A. Infect Immun 69:2487–2493

Ciprian A, Pijoan C, Cruz T, Camacho J, Tortora J, Colmenares G, Lopez-Revilla R, de la Garza M (1988) *Mycoplasma hyopneumoniae* increases the susceptibility of pigs to experimental *Pasteurella multocida* pneumonia. Can J Vet Res 52:434–438

Collins FM, Niederbuhl CJ, Campbell SG (1983) Bactericidal activity of alveolar and peritoneal macrophages exposed in vitro to three strains of *Pasteurella multocida*. Infect Immun 39:779–784

Confer AW (2009) Update on bacterial pathogenesis in BRD. Anim Health Res Rev 10:145–148

Cox AJ, Hunt ML, Boyce JD, Adler B (2003) Functional characterization of HgbB, a new hemoglobin binding protein of *Pasteurella multocida*. Microb Pathog 34:287–296

Curtis PE, Ollerhead GE (1981) Investigation to determine whether healthy chickens and turkeys are oral carriers of *Pasteurella multocida*. Vet Rec 108:206–207

Dabo S, Confer A, Quijano-Blas R (2003) Molecular and immunological characterization of *Pasteurella multocida* serotype A:3 OmpA: evidence of its role in *P. multocida* interaction with extracellular matrix molecules. Microb Pathog 35:147–157

Dabo SM, Confer AW, Hartson SD (2005) Adherence of *Pasteurella multocida* to fibronectin. Vet Microbiol 110:265–275

Dabo SM, Taylor JD, Confer AW (2008) *Pasteurella multocida* and bovine respiratory disease. Anim Health Res Rev 8:129–150

De Alwis MCL (1999) ACIAR monograph 57: haemorrhagic septicaemia. Australian Centre for International Agricultural Research, Canberra

De Alwis MCL, Vipulasiri AA (1980) An epizootiological study of haemorrhagic septicaemia in Sri Lanka. Ceylon Vet J 28:24–35

De Alwis MCL, Wijewardana TG, Sivaram A, Vipulasiri AA (1986) The carrier and antibody status of cattle and buffaloes exposed to haemorrhagic septicaemia: investigations on survivors following natural outbreaks. Sri Lankan Vet J 34:33–42

Deeb BJ, DiGiacomo RF, Bernard BL, Silbernagel SM (1990) *Pasteurella multocida* and *Bordetella bronchiseptica* infections in rabbits. J Clin Microbiol 28:70–75

Derieux WT (1978) Response of young chickens and turkeys to virulent and avirulent *Pasteurella multocida* administered by various routes. Avian Dis 22:131–139

Despande MS, Ambagala TC, Ambagala APN, Kehrli ME, Srikumaran S (2002) Bovine CD18 is necessary and sufficient to mediate *Mannheimia* (*Pasteurella*) *haemolytica* leukotoxin-induced cytolysis. Infect Immun 70:5058–5064

DiGiacomo RF, Deeb BJ, Giddens WE, Bernard BL, Chengappa MM (1989) Atrophic rhinitis in New Zealand white rabbits infected with *Pasteurella multocida*. Am J Vet Res 50:1460–1465

DiGiacomo RF, Xu YM, Allen V, Hunter MH, Pearson GR (1991) Naturally-acquired *Pasteurella multocida* infection in rabbits: clinicopathological aspects. Can J Vet Res 55:234–238

Dillehay DL, Paul KS, DiGiacomo RF, Chengappa MM (1991) Pathogenicity of *Pasteurella multocida* A3 in Flemish Giant and New Zealand White rabbits. Lab Anim Sci 25:337–341

Dominick MA, Rimler RB (1988) Turbinate osteoporosis in pigs following intranasal inoculation of purified *Pasteurella* toxin: histomorphometric and ultrastructural studies. Vet Pathol 25:17–27

Dowling AJ, Hodgson C, Schock A, Donachie W, Ekersall PD, McKendrick IJ (2002) Experimental induction of pneumonic pasteurellosis in calves by intratracheal infection with *Pasteurella multocida* biotype A:3. Res Vet Sci 73:37–44

Eliás B, Boros G, Albert M, Tuboly S, Gergely P, Papp L, Barna Vetró I, Rafai P, Molnár E (1990) Clinical and pathological effects of the dermonecrotic toxin of *Bordetella bronchiseptica* and *Pasteurella multocida* in specific-pathogen-free piglets. Nihon Juigaku Zasshi 52:677–688

Ewers C, Lubke-Becker A, Bethe A, Kiebling S, Filter M, Wieler LH (2006) Virulence genotype of *Pasteurella multocida* strains isolated from different hosts with various disease status. Vet Microbiol 114:304–317

Freshwater A (2008) Why your housecat's trite little bite could cause you quite a fright: a study of domestic felines on the occurrence and antibiotic susceptibility of *Pasteurella multocida*. Zoonoses Public Health 55:507–513

Fuller TE, Kennedy MJ, Lowery DE (2000) Identification of *Pasteurella multocida* virulence genes in a septicemic mouse model using signature-tagged mutagenesis. Microb Pathog 29:25–38

Fulton RW (2009) Bovine respiratory disease research (1983–2009). Anim Health Res Rev 10:131–139

Galdiero M, De Martino L, Pagnini U, Pisciotta M-G, Galdiero E (2001) Interactions between bovine endothelial cells and *Pasteurella multocida*: association and invasion. Res Microbiol 152:57–65

Glass SE, Panigrahy B (1990) Dermal necrosis caused by *Pasteurella multocida* infection in turkeys. Avian Dis 34:491–494

Hall WF, Bane DP, Kilroy CR, Essex-Sorlie DL (1990) A model for the induction of *Pasteurella multocida* type-A pneumonia in pigs. Can J Vet Res 54:238–243

Hamilton TDC, Roe JM, Hayes CM, Webster AJF (1998) Effect of ammonia inhalation and acetic acid pretreatment on colonisation kinetics of toxigenic *Pasteurella multocida* within the upper respiratory tract of swine. J Clin Microbiol 36:1260–1265

Hansen MS, Pors SE, Jensen HE, Bille-Jensen V, Bisgaard M, Flachs EM, Nielsen OL (2010) An investigation of the pathology and pathogens associated with porcine respiratory disease complex in Denmark. J Comp Path 143:120–131

Harel J, Cote S, Jacques M (1990) Restriction endonuclease analysis of porcine *Pasteurella multocida* isolates from Quebec. Can J Vet Res 54:422–426

Harmon BG, Glisson JR, Nunnally JC (1992) Turkey macrophage and heterophil bactericidal activity against *Pasteurella multocida*. Avian Dis 36:989–991

Harper M, Boyce JD, Wilkie IW, Adler B (2003) Signature-tagged mutagenesis of *Pasteurella multocida* identifies mutants displaying differential virulence characteristics in mice and chickens. Infect Immun 71:5440–5446

Harper M, Boyce JD, Adler B (2006) *Pasteurella multocida* pathogenesis: 125 years after Pasteur. FEMS Microbiol Lett 265:1–10

Hassanin HH, Toth TE, Eldimerdash M, Siegel PB (1995) Stimulation of avian respiratory phagocytes by *Pasteurella multocida*—effects of the route of exposure, bacterial dosage and strain, and the age of chickens. Vet Microbiol 46:401–413

Hindman LJ, Harvey WF, Costanzo GR, Converse KA, Stein G (1997) Avian cholera in ospreys: first occurrence and possible mode of transmission. J Field Ornithol 68:503–508

Horadagoda NU, Hodgson JC, Moon GM, Wijewardana TG, Eckersall PD (2001) Role of endotoxin in the pathogenesis of haemorrhagic septicaemia in the buffalo. Microb Pathog 30:171–178

Hundt MJ, Ruffolo CG (2005) Interaction of *Pasteurella multocida* with free-living amoebae. Appl Environ Microbiol 71:5458–5464

Hunter B, Wobeser G (1980) Pathology of experimental avian cholera in mallard ducks. Avian Dis 24:403–414

Jordan RW, Roe JM (2004) An experimental mouse model of progressive atrophic rhinitis of swine. Vet Microbiol 103:201–207

Jubb KV, Maxie MG (2007) Pathology of domestic animals, 5th edn. Elsevier Saunders, New York

Kachlany SC, Planet PJ, DeSalle R, Fine DH, Figurski DH, Kaplan JB (2001) flp-1, the first representative of a new pilin gene subfamily, is required for non-specific adherence of *Actinobacillus actinomycetemcomitans*. Mol Microbiol 40:542–554

Kaden V, Teifke JP, Polster V (2001) Progressive atrophic rhinitis—a rare disease in wild boar. Zeitschr Jagdenwissensch 47:17–25

Lee MD, Wooley RE, Glisson JR (1994) Invasion of epithelial cell monolayers by turkey strains of *Pasteurella multocida*. Avian Dis 38:72–77

Mackie JT, Barton M, Kettlewell J (1992) *Pasteurella multocida* septicaemia in pigs. Aust Vet J 69:227–228

Martineau-Doize B, Dumas G, Larochelle R, Frantz JC, Martineau G-P (1991) Atrophic rhinitis caused by *Pasteurella multocida* type D: morphometric analysis. Can J Vet Res 55:224–228

Matsumoto M, Strain JG, Engel HN (1991) The fate of *Pasteurella multocida* after intratracheal inoculation into turkeys. Poult Sci 70:2259–2266

May BJ, Zhang Q, Li LL, Paustian ML, Whittam TS, Kapur V (2001) Complete genomic sequence of *Pasteurella multocida*, Pm70. Proc Natl Acad Sci USA 98:3460–3465

Mbuthia PG, Njagi LW, Nyaga PN, Bebora LC, Minga U, Kamundia J, Olsen JE (2008) *Pasteurella multocida* in scavenging family chickens and ducks: carrier status, age susceptibility and transmission between species. Avian Pathol 37:51–57

McFadden MJ, Christensen H, Fairley AR, Hill FI, Gill JM, Keeling SE, Spence RP (2011) Outbreaks of pleuritis and peritonitis in calves associated with *Pasteurella multocida* capsular type B strain. N Z Vet J 59:40–45

Morrison RB, Hilley HD, Leman AD (1985) The association between pneumonia and atrophic rhinitis in slaughter weight swine. Can Vet J 26:95–97

Morton DJ, Whitby PW, Jin H, Ren Z, Stull TL (1999) Effect of multiple mutations in the hemoglobin- and hemoglobin-haptoglobin-binding proteins, HgpA, HgpB, and HgpC, of *Haemophilus influenzae* type b. Infect Immun 67:2729–2739

Muhairwa AP, Christensen JP, Bisgaaard M (2000) Investigations on the carrier rate of *Pasteurella multocida* in healthy commercial poultry flocks and flocks affected by fowl cholera. Avian Pathol 29:133–142

Mullan PB, Lax AJ (1998) *Pasteurella multocida* toxin stimulates bone resorption by osteoclasts via interaction with osteoblasts. Calcif Tiss Int 63:340–345

Mullen LM, Bossé JT, Nair SP, Ward JM, Rycroft AN, Robertson G, Langford PR, Henderson B (2008a) Pasteurellaceae ComE1 proteins combine the properties of fibronectin adhesins and DNA binding competence proteins. PLoS ONE 3:e3991

Mullen LM, Nair SP, Ward JM, Rycroft AN, Williams RJ, Robertson G, Mordan NJ, Henderson B (2008b) Novel adhesin from *Pasteurella multocida* that binds to the integrin-binding fibronectin FnIII9-10 repeats. Infect Immun 76:1093–1104

Muller G, Kohler H (1997) Experimental affecting of pulmonary clearance of *Pasteurella multocida* induced pneumonia in swine. Berliner Munchener Tierarztl Wochenschr 110:378–380

Ogunnariwo JA, Schryvers AB (2001) Characterization of a novel transferrin receptor in bovine strains of *Pasteurella multocida*. J Bacteriol 183:890–1084

Pabs-Garnon LF, Soltys MA (1971) Methods of transmission of fowl cholera in turkeys. Am J Vet Res 32:1119–1120

Paisley LG, Vraa-Andersen L, Dybkjaer L, Møller K, Christensen G, Mousing J, Agger JF (1993) An epidemiologic and economic study of respiratory diseases in two conventional Danish swine herds. II. associations between lesions present at slaughter and mean daily gains during specific intervals of the growth period. Acta Vet Scand 34:331–344

Paustian ML, May BJ, Kapur V (2001) *Pasteurella multocida* gene expression in response to iron limitation. Infect Immun 69:4109–4115

Paustian ML, May BJ, Cao D, Boley D, Kapur V (2002) Transcriptional response of *Pasteurella multocida* to defined iron sources. J Bacteriol 184:6714–6720

Pelicic V (2008) Type IV pili: e pluribus unum? Mol Microbiol 68:827–837

Petersen KD, Christensen JP, Permin A, Bisgard M (2001) Virulence of *Pasteurella multocida* subsp multocida isolated from outbreaks of avian cholera in wild birds for domestic poultry and game birds. Avian Pathol 30:27–31

Pijoan C, Trigo F (1990) Bacterial adhesion to mucosal surfaces with special reference to *Pasteurella multocida* isolates from atrophic rhinitis. Can J Vet Res 54(Suppl):S16–S21

Poermadjaja B, Frost A (2000) Phagocytic uptake and killing of virulent and avirulent strains of *Pasteurella multocida* of capsular serotype A by chicken macrophages. Vet Microbiol 72:163–171

Pruimboom IM, Rimler RB, Ackermann MR (1999) Enhanced adhesion of *Pasteurella multocida* to cultured turkey peripheral blood monocytes. Infect Immun 67:1292–1296

Purdy CW, Straus DC, Chirase N, Ayers JR, Hoover MD (2003) Effects of aerosolized dust in goats on lung clearance of *Pasteurella* and *Mannheimia* species. Curr Microbiol 46:174–179

Rad MA, Movassaghi R, Sharifi K, Naseri Z, Seifi HA (2011) Two outbreaks of *Pasteurella multocida* septicaemia in neonatal lambs. Comp Clin Pathol 20:57–59

Ramdani (1992) Studies on pathogenesis and immunochemistry of *Pasteurella multocida* model infection in mice. PhD thesis, Monash University, Melbourne

Rhoades KR, Rimler RB (1990) *Pasteurella multocida* colonization and invasion in experimentally exposed turkey poults. Avian Dis 34:381–383

Ross RF (2006) *Pasteurella multocida* and its role in porcine pneumonia. Anim Health Res Rev 7:13–29

Rouman MP, Van de Goor PT (1988) Atrophic rhinitis in cattle. Tijdschr Diergeneesk 113:28

Ruffolo CG, Tennent JM, Michalski WP, Adler B (1997) Identification, purification, and characterization of the type 4 fimbriae of *Pasteurella multocida*. Infect Immun 65:339–343

Saif YM (2008) Diseases of poultry. Wiley-Blackwell, Ames

Scheidt A, Mayrose VB, Hill MA, Clark LK, Cline TR, Knox KE, Runnels LJ, Frantz S, Einstein ME (1990) Relationship of growth performance to pneumonia and atrophic rhinitis detected in pigs at slaughter. J Am Vet Med Assoc 196:881–884

Sheikh MA, Yaqoob T, Baig MS, Mahamood AM, Shakoori AR (1994) The epidemiology of haemorrhagic septicaemia of buffaloes in Pakistan. Buffalo J 10:229–236

Talan DA, Citron DM, Abrahamian FM, Moran GJ, Goldstein EJC (1999) Bacteriologic analysis of infected dog and cat bites. New Eng J Med 340:85–92

Tatum F, Yersin A, Briggs R (2005) Construction and virulence of a *Pasteurella multocida fhaB2* mutant in turkeys. Microb Pathog 39:9–17

Tatum FM, Tabatabai LB, Briggs RE (2009) Sialic acid uptake is necessary for virulence of *Pasteurella multocida* in turkeys. Microb Pathog 46:337–344

Tennent JM, Mattick JS (1994) Type 4 fimbriae. In: Klemm P (ed) Fimbriae: adhesion, genetics, biogenesis and vaccines. CRC Press Inc., Boca Raton

Tsuji M, Matsumoto M (1990) Immune defence mechanism against blood-borne *Pasteurella multocida* in turkeys. Res Vet Sci 48:344–349

Wassenaar TM, Gaastra W (2001) Bacterial virulence: can we draw the line? FEMS Microbiol Lett 201:1–7

Watson PJ, Davies RL (2002) Outbreak of *Pasteurella multocida* septicaemia in neonatal lambs. Vet Rec 151:420–422

Wilkie IW, Grimes SE, O'Boyle D, Frost AJ (2000) The virulence and protective efficacy for chickens of *Pasteurella multocida* administered by different routes. Vet Microbiol 72:157–168

Williams ES, Runde DE, Mills K, Holler LD (1987) Avian cholera in a gyrfalcon (*Falco rusticolus*). Avian Dis 31:380–382

Zhao G, Pijoan C, Murtaugh MP (1993) Epidemiology of *Pasteurella multocida* in a farrow to finish swine herd. Can J Vet Res 57:136–138

Pathogenomics of *Pasteurella multocida*

J. D. Boyce, T. Seemann, B. Adler and M. Harper

Abstract The first complete genome sequence of the *P. multocida* avian isolate Pm70 was reported in 2001. Analysis of the genome identified many predicted virulence genes, including two encoding homologues of the *Bordetella pertussis* filamentous haemagluttinins, and genes involved in iron transport and metabolism. Availability of the genome sequence allowed for a range of whole-genome transcriptomic and proteomic analyses and these have helped us understand how *P. multocida* responds to growth in the presence of antibiotics, under low iron conditions and in the host. Unfortunately, no new *P. multocida* genome sequences were determined during the rest of the decade, limiting any possible comparative genomic analyses until recently, when several new genome sequences have become available. Here we use the available data to identify a number of important similarities and differences between the strains and determine their phylogenetic relationships. Interestingly, based on the current data there is no clear correlation between phylogenetic relatedness and host predilection or disease.

Contents

J. D. Boyce · B. Adler (✉) · M. Harper
Australian Research Council Centre of Excellence in Structural and Functional Microbial Genomics, Department of Microbiology, Monash University,
Clayton, Victoria 3800, Australia
e-mail: ben.adler@monash.edu

T. Seemann
Victorian Bioinformatics Consortium, Department of Microbiology,
Monash University, Clayton, Victoria 3800, Australia

Current Topics in Microbiology and Immunology (2012) 361: 23–38
DOI: 10.1007/82_2012_203

Published Online: 9 March 2012

1 Sequenced *P. multocida* Genomes

There are currently two fully sequenced and annotated *P. multocida* genomes available: *P. multocida* strain Pm70 (accession number AE004439) (May et al. 2001) and *P. multocida* strain 36950 (accession number CP003022) (Michael et al. 2012b). Strain Pm70 is an avian isolate and while it was initially reported as a capsular type A (hyaluronan capsule), Heddleston serovar 3 (A:3) strain (May et al. 2001) analysis of the capsule biosynthesis locus indicates that it produces a type F (chondroitin) capsule, as it lacks genes for hyaluronan synthesis (Townsend et al. 2001), but encodes a chondroitin synthase. Strain 36950 is an A:3 strain isolated from a calf with bovine respiratory disease (BRD). This strain is resistant to a range of antibiotics, including tetracyclines, chloramphenicol, sulphonamides, spectinomycin, streptomycin, enrofloxacin, florfenicol, tilmicosin and tulathromycin (Michael et al. 2012b). Three other *P. multocida* genomes are also publically available, but they are incomplete, limiting the range of analyses that can be undertaken. These are strain P3480, isolated from a pneumonic pig lung, strain Anand1 from a goat (Anand1G) and *P. multocida* subps. *gallicida* strain Anand1 from poultry (Anand1P; Ahir et al. 2011). In addition, we have recently determined incomplete genome sequences (using Illumina GAIIx technology) of the two fowl cholera isolates X73 and VP161 (both A:1), the bovine haemorrhagic septicaemia isolate M1404 (B:2) and the swine isolate P903 (D:11).

All of the currently available *P. multocida* genomes are between 2 and 2.4 Mbp in length and comprise a single circular genome with a G+C content of between 40 and 41% (Table 1). Analysis of the Pm70 genome identified more than 100 genes predicted to be involved in virulence; of particular note were two very large genes (7.8 and 11.8 Kb in length) encoding predicted homologues of the *Bordetella pertussis* filamentous haemagglutinins (May et al. 2001). These have since been defined as important *P. multocida* virulence factors (Tatum et al. 2005). The Pm70 genome also contains the complete set of genes required for the following pathways: TCA cycle, glycolysis, glyconeogenesis, oxidative pentose phosphate and Entner–Doudoroff (May et al. 2001).

Table 1 Genome features of sequenced *P. multocida* strains

Strain	Complete	Host/tissue	Genome size (Mbp)	G+C content	CDS[a]
Pm70	Yes	Chicken	2.26	40.40	2,089
36950	Yes	Bovine/lung	2.35	40.44	2,202
Anand1G	No	Caprine	2.29	40.52	2,319
Anand1P	No	Avian/liver	2.02	40.22	2,337
P3480	No	Porcine/lung	2.08	40.15	1,933
X73	No	Avian	2.31	40.31	2,130
VP161	No	Avian	2.26	40.21	2,085
P903	No	Swine	2.35	40.14	2,184
M1404	No	Bovine	2.30	40.27	2,136

[a] CDS = predicted number of coding sequences

2 Comparison of the Complete Genomes of Pm70 and 36950

Currently only the Pm70 and 36950 genomes are closed and complete. Both genomes contain six rRNA operons comprising 20 and 19 genes, respectively. Comparison of the sequences shows that they are highly colinear and contain nine large, locally colinear blocks (LCB) (Fig. 1). However, there are areas of novel sequence identified in each genome. Strain 36950 contains a large integrative conjugative element designated ICE*Pmu1*, which is not found in Pm70 (approximately region 270–360 Kb; Fig. 1). This 82 Kb element contains approximately 88 genes and is capable of conjugative movement into *P. multocida*, *Mannheimia haemolytica* and *Escherichia coli*, as it contains the genes required for integration/excision and conjugation (Michael et al. 2012b). The element also contains genes encoding resistance to streptomycin, spectinomycin, tetracycline, chloramphenicol/florfenicol, sulphonamides, tilmicosin/clindamycin and tulathromycin, which explains the broad antibiotic resistances exhibited by this strain. Strain 36950 also contains point mutations outside the ICE*Pmu1* element within *gyrA* and *parC*, potentially altering the target sites of quinolone/fluoroquinolone action (Michael et al. 2012a). The 36950 strain also contains genes for uptake/metabolism of mannose, trehalose and xylose, which are not present in Pm70.

The Pm70 genome also contains a small number of regions not found in 36950. Most of the Pm70 region encoding the genes *pm1935–pm1949* (approximately region 2,172–2,187 Kb; Fig. 1) is not present in strain 36950. This section of DNA has a low G+C content (37.6%) and *pm1949* encodes a predicted site-specific recombinase, suggesting that this region of the chromosome may have been laterally acquired. Pm70 also contains two genes (*pm0698–pm0699*) encoding components of a predicted restriction/modification system which is not present in strain 36950 and two unique phage-like genes (*pm0202* and *pm0203*).

A comparison of the strain 36950 genome with Pm70 shows that strain 36950 contains more than 30 non-functional genes (pseudogenes). These include *pfhB1*, encoding one of the two Pm70 filamentous haemagglutinins, *hsf_1*, encoding one of the two predicted autotransporter surface adhesins, four genes (*pfhR*, *pm0745*, and

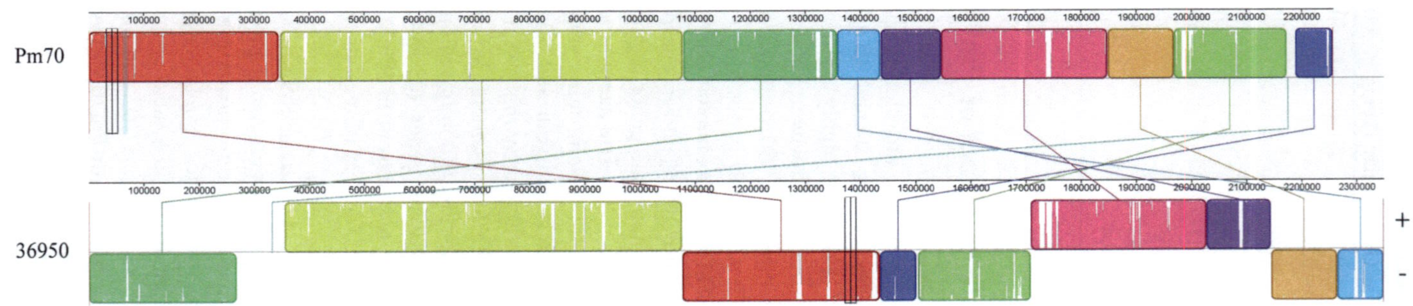

Fig. 1 Comparison of the Pm70 and 36950 closed genome sequences using progressiveMauve alignment (Darling et al. 2010). Rectangles of similar colour show colinear blocks of genes. In the 36950 panel, *the upper coloured blocks* labelled "+" indicate genes oriented in the same direction as in the Pm70 genome while blocks labelled "−" indicate genes oriented in the reverse direction to the Pm70 genome. Areas of low identity within colinear blocks are shown by reduced height of the shading

two homologues of *pm1622*) encoding predicted Ton-B dependent outer membrane receptors and *plpE*, encoding a protective outer membrane lipoprotein. In addition, Pm70 expresses three highly related, predicted extracellular solute iron binding proteins (Afu1, Afu2 and Afu3) and although the corresponding genes are present in strain 36950, both *afu2* and *afu3* are present as pseudogenes. As a number of these pseudogenes have putative functional paralogues (PfhB1 and PfhB2, Hsf1 and Hsf2), and there is likely significant redundancy in the expressed iron transporters, it is possible that the loss of these functional genes has little impact on the phenotype of this strain.

3 Analysis of the Incomplete Genome Sequences of Strains X73, VP161, Anand1P, Anand1G, P903, M1404 and P3480

Comparison of the Pm70 and 36950 genomes with the incomplete genome sequences of strains X73, VP161, Anand1P, Anand1G, P903, M1404 and P3480 using BLAST, identified a number of regions that were absent, or highly divergent, in all of the comparison strains (Fig. 2). The ICE*Pmu1* element is not present in any of the sequenced strains except 36950; this absence correlates with the low level of antibiotic resistance observed in the other strains. Similarly, the Pm70 genes *pm1935–pm1949* are not present in any of the other genomes. These comparisons also suggest that the sequence coverage of P3480 is low, as there are two very large regions of the genome that are missing from the sequence data obtained for this strain (Fig. 2a; 1400–1500 Kb and 1760–1970 Kb on the Pm70 map).

Comparison of the genomes identified large sections of colinear genes in all genomes, with the largest colinear block approximately 250 Kb in length and containing more than 200 genes (Pm70 region *pm0506–pm0712*). Progressive Mauve alignment (Darling et al. 2010) was used to determine colinear orthologues between genomes at a 60% identity level with greater than 70% gene coverage. Across all nine genomes there is a shared set of 1,100 genes and each of the strains contains between 73 (Pm70) and 638 (Anand1P) unique genes (Fig. 3a). It is likely that the number of core genes identified by this analysis is artificially low due to the low coverage of the P3480 sequence and the high number of incomplete and broken gene sequences in the Anand1P and Anand1G genomes. The number of unique genes may also be overestimated in the incomplete genomes due to broken gene sequences. Therefore, we repeated the orthologue analysis with only the high quality and high coverage genomes, Pm70, 36950, X73, P903 and M1404. Across this set of five strains, there is a shared set of more than 1,780 genes, or 88% of the Pm70 gene content (Fig. 3b) and the pan genome, or total gene set, is more than 2,800 genes. Pm70 contains approximately 90 unique genes; strain 36950, 170 genes; X73, 177 genes; P903, 261 genes and M1404, 215 genes. Of the 90 unique Pm70 genes, more than 80% encode hypothetical proteins with no, or poorly defined, predicted function. This very high percentage of hypothetical proteins in the unique gene set is also observed in the other strains. For 36950, more than 80

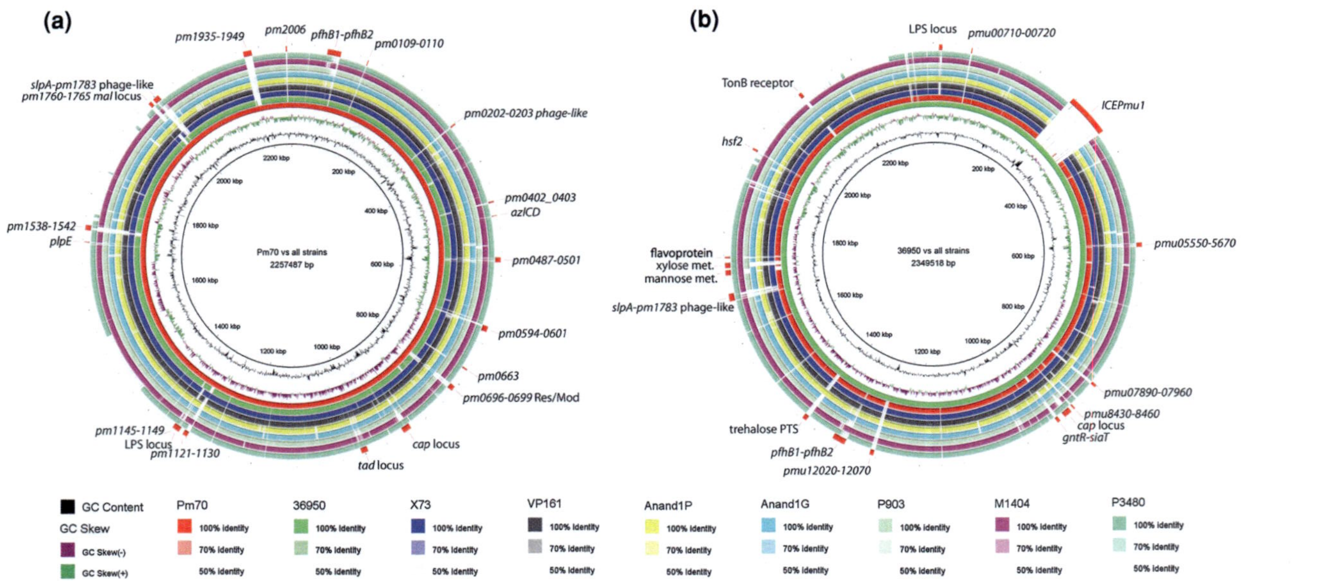

Fig. 2 Comparison of the genomes of 36950, X73, VP161, Anand1P, Anand1G, P903, M1404 and P3480 with the genome of Pm70 (**a**) and Pm70, X73, VP161, Anand1P, Anand1G, P903, M1404 and P3480 with the genome of 36950 (**b**). The three *inner rings* show the DNA size, GC content and GC skew of the reference genome (Pm70 in **a** and 36950 in **b**). The nine *outer rings* show regions of the comparison genomes which match the reference genome (shown as the inside *solid coloured ring*). *Red arcs* on the outside identify particular regions that are absent, or show reduced identity, in some or all of the comparison genomes. Figures were drawn using BLAST ring image generator (Alikhan et al. 2011)

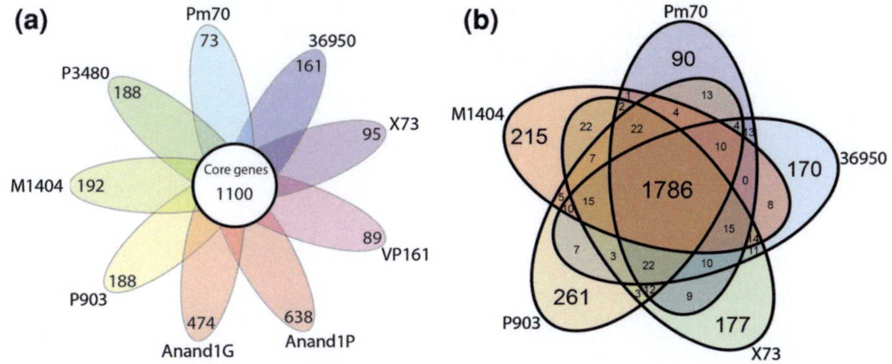

Fig. 3 Flowerplot of the unique genes in each of the *P. multocida* strain Pm70, 36950, X73, VP161, Anand1P, Anand1G, P903, M1404 and P3480 genomes (**a**) and a five set Venn diagram showing unique genes and all combinations of shared genes in the strains Pm70, 36950, X73, P903 and M1404. Orthologues were identified using ProgressiveMauve alignment (Darling et al. 2010) and have not been manually curated

of the 170 unique genes are within the ICE*Pmu1* element, whereas for M1404 approximately 60 of the 215 unique genes are on DNA sections predicted to be within temperate phage or phage-derived elements.

The phylogenetic relationship between the nine strains was predicted by comparison of all single nucleotide polymorphisms at conserved genome positions. For the overall phylogenetic analysis, the three related species, *Gallibacterium anatis*, *Mannheimia haemolytica* and *P. dagmatis* were included as outliers (Fig. 4a). This tree, based on 7,931 SNPs at shared positions in all 12 strains, shows that *P. dagmatis* is the most closely related species and all of the *P. multocida* strains are very closely related to each other. The finer phylogenetic relatedness of the nine *P. multocida* strains shows that there appears to be little or no correlation between the phylogenic relatedness of the strains and the country of isolation, the serogroup or serovar, the disease or the host predilection (Fig. 4b). It will be of interest to discover whether this trend continues as more *P. multocida* genomes are sequenced. A similar phylogeny was produced using the nucleotide sequences of the seven multi-locus sequencing typing genes described previously by Subaaharan et al. (2010) (data not shown).

4 Plasmids Carried by *P. multocida* Strains

While Pm70 and 36950 do not contain plasmids, some *P. multocida* strains carry multiple plasmids which may be either cryptic or carry antibiotic resistance genes. Initial studies on a *P. multocida* type D strain, isolated from a pig with atrophic rhinitis, identified the presence of four plasmids ranging in size from 1.5 to 5.4 Kb (Wright et al. 1997). The largest of these plasmids, designated pIG1, could replicate in both *P. multocida* and *E. coli* and carried genes for sulphonamide and streptomycin resistance. This plasmid was further identified in 11 of 12 Australian *P. multocida* pig isolates. The replication region of pIG1 was similar to the

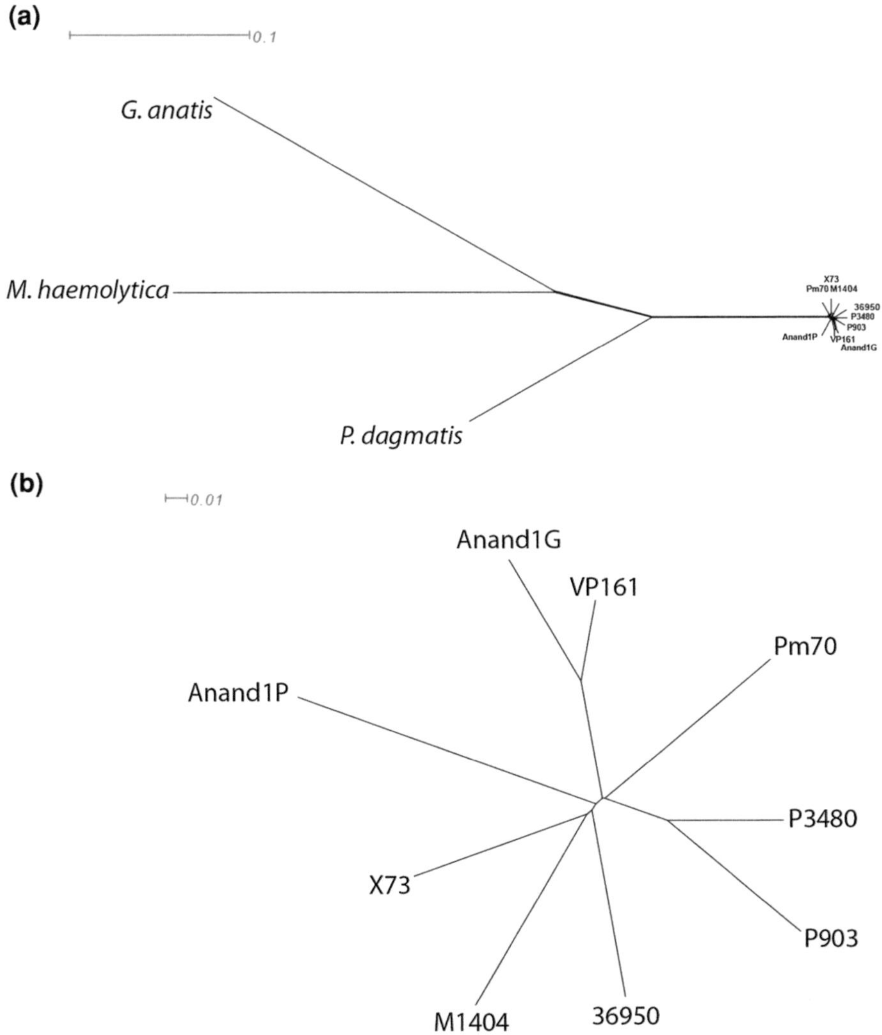

Fig. 4 Unrooted neighbour-joining tree showing the phylogenetic relationship between *Gallibacterium anatis*, *Mannheimia haemolytica*, *Pasteurella dagmatis* and the *P. multocida* strains Pm70, 36950, X73, VP161, Anand1P, Anand1G, P903, M1404 and P3480 (**a**) and only the *P. multocida* strains Pm70, 36950, X73, VP161, Anand1P, Anand1G, P903, M1404 and P3480 (**b**). Phylogenetic relatedness was determined by analysis of only the single nucleotide polymorphisms (SNPs) found at conserved positions in all genomes of the comparison set. The tree in (**a**) is based on analysis of 7,931 SNPs while the tree in (**b**) is based on analysis of 54,415 SNPs. The tree was rendered using SplitsTree4 (Huson and Bryant 2006)

corresponding region in the *Haemophilus ducreyi* plasmid, pLS88, and the *Mannheimia [P.] haemolytica* plasmid, pYFC1. Mobilisation of pIG1 into *E. coli* SM10 was dependent on three genes with shared identity to the *E. coli* ColE1 plasmid mobilisation genes *mbeA*, *mbeB* and *mbeC* (Wright et al. 1997).

A number of *P. multocida* plasmids have now been isolated that carry multiple resistance genes. The plasmid pVM111, isolated from a fowl cholera-causing strain, carries genes for sulphonamide, streptomycin and tetracycline resistance (Kehrenberg et al. 2003), while the mobilisable plasmid pCCK381 that was isolated from a serogroup A bovine respiratory disease isolate, carries genes for florfenicol, chloramphenicol and sulphonamide resistance. Interestingly, the plasmid replication and mobilisation genes on pCCK381 were highly similar to those on the *Dichelobacter nodosus* plasmid pDN1, whereas other sections were nearly identical to the *E. coli* plasmid pMBSF1 and the *Vibrio salmonicida* plasmid pRVS1. These data suggest that pCCK381 is the result of interplasmid recombination which may have been selected during florfenicol treatment of infected livestock (Kehrenberg and Schwarz 2005).

Recently it was shown that some antibiotic resistant *P. multocida* strains carry multiple plasmids, each with one or two antibiotic resistance genes (San Millan et al. 2009). While studying ampicillin resistant and multiple antibiotic resistant *P. multocida* isolates from pigs, a range of highly related plasmids was identified; pB1000 which was a ColE1-type plasmid, 4,613 bp in length and carried the ampicillin resistance gene $bla_{\mathrm{ROB\text{-}1}}$, pB1005 encoding streptomycin and sulphonamide resistance and pB1006, pB1001 and p9956 that carried the tetracycline resistance genes *tet*(O), *tet*(B) or *tet*(H), respectively. Some strains contained more than one of these plasmids; ampicillin, streptomycin and sulphonamide resistant strains generally contained both pB1000 and pB1005, whereas some ampicillin, streptomycin, sulphonamide and tetracycline resistant strains contained pB1000, pB1005 and pB1006. The plasmids all contained highly related mobilisation and *oriV* regions. The *oriV* region is predicted to be responsible for plasmid replication and copy number control. Although the *oriV* regions of the different plasmids were closely related, they were not identical; it is assumed that these differences are sufficient to allow stable coexistence of up to at least three of these plasmids in the same cell (San Millan et al. 2009).

5 Bacteriophages of *P. multocida*

The genome sequence of Pm70 contains very few phage-like sequences or insertion elements and does not appear to contain any complete temperate phage genomes. However, the region between *slpA* and *pm1783* contains a number of hypothetical genes, two of which encode proteins with similarity to phage proteins: a predicted phage integrase (SlpA) and a predicted subunit of a phage terminase (Pm1777). This set of genes is also present in strain 36950, but not in X73, VP161, P903 or M1404 (Fig. 3). The Pm70 genes *pm0282* and *pm0283* are present only in Pm70 and encode proteins with similarity to the *E. coli* phage lambda proteins ea59 and ea31, respectively.

Some *P. multocida* strains do contain complete temperate phage genomes (Campoy et al. 2006; Nielsen and Rosdahl 1990; Pullinger et al. 2004). Importantly, the primary virulence factor for porcine atrophic rhinitis, the *P. multocida* PMT toxin, is carried on a lysogenic bacteriophage (Pullinger et al. 2004). Three toxigenic serotype D strains (LFB3, P27 and G4) were treated with the DNA-damaging agent mitomycin C and all released isometric-headed bacteriophage of the *Siphoviridae* class; this class also includes lambda-like phage. The *toxA* gene, encoding the *P. multocida* PMT toxin, was present in the genomes of all phage examined. The presence of *toxA* on a lysogenic phage may be important for both the expression and release of the toxin during disease progression. PMT has no signal sequence and is not secreted from intact *P. multocida* cells (iDali et al. 1991; Pullinger et al. 2004). However, PMT must be released from the bacteria to allow it to interact with the cell surface receptors on host cells and it has been suggested that induction of the phage lytic cycle during host infection may result in bacterial lysis and PMT release. Furthermore, the amplification of the phage genome during lytic growth may increase the expression of the PMT under these conditions (Pullinger et al. 2004). The details of the mode of action of PMT are reported in Molecular Biology of *Pasteurella Multocida* toxin.

One temperate transducing phage (F108) has been isolated from a *P. multocida* serovar A strain and its genome completely sequenced (Campoy et al. 2006). The morphology of the F108 temperate phage is similar to phage belonging to the *Myoviridae* family, with a 50 nm hexagonal head and a 120 nm long tail. Its genome is approximately 30 Kb in length, predicted to have cohesive ends and contains more than 40 genes. Neither F108, nor the PMT-containing phages, have been identified in any of the currently sequenced *P. multocida* genomes.

Analysis of the genomes of 36950, X73, VP161, P903 and M1404 identified a small number of putative temperate phage, or temperate phage-derived genes. Both strains X73 and M1404 contain genes encoding proteins with identity to Mu-like bacteriophage proteins; in M1404 this section is greater than 25 Kb in length and contains more than 40 genes, while in X73 it is approximately 14 Kb in length and contains 14 genes. While both sections have identity with Mu-like phages, there is little shared identity, indicating that these sections are not the result of the acquisition of an identical phage. Strain M1404 also contains a novel 15.7 Kb region containing genes with identity to known bacteriophage genes. This section comprises 19 ORFS and is situated at the $tRNA^{Met}$ downstream of the Pm70 gene *pm1654*; it is of note that the F108 phage specifically integrates into the $t33tRNA^{leu}$ (Campoy et al. 2006) and the lysogenic phage which carries the PMT toxin preferentially integrates into the $t3tRNA^{leu}$.

6 Whole-Genome Transcriptomic Analyses

The completion of the Pm70 genome sequence (May et al. 2001) facilitated the construction of *P. multocida* DNA microarrays containing all 2,015 genes identified in this genome. These microarrays were essential for a range of new analyses into *P. multocida* pathogenic mechanisms.

Iron is an essential nutrient for most bacteria, but free iron is present at limiting concentrations in the host. Indeed, in mammalian and avian hosts iron is primarily bound to carrier proteins, including transferrin, lactoferrin and ferritin (Paustian et al. 2002). Initial annotation of the Pm70 genome indicated that it contained at least 53 genes encoding proteins predicted to be involved in iron uptake or acquisition (May et al. 2001); current analysis suggests that this number is more than 60 genes. Thus, it is likely that the acquisition of iron is critical for *P. multocida* survival in vivo. The large number of genes involved in iron uptake also suggests that *P. multocida* has a number of redundant, or partially redundant, iron uptake systems.

A number of DNA microarray studies has analysed the whole-genome transcriptional response of *P. multocida* to growth under limiting concentrations of iron or in the presence of different sources of iron (May et al. 2001; Paustian et al. 2002, 2001). When *P. multocida* strain Pm70 was grown under iron limitation in an otherwise rich medium (brain heart infusion containing 200 mM 2,2'-dipyridyl), more than 150 genes showed significantly (more than two-fold) altered expression. In general, the response to iron limitation resulted in a decrease in expression of genes involved in energy metabolism and an increase in expression of genes involved in iron transport, amino acid metabolism and central intermediary metabolism (Paustian et al. 2001). The genes *eno*, *gapdh*, and *pgk*, which encode proteins of the glycolytic pathway, were all down regulated. As expected under iron-limited conditions, major increases in expression were observed for numerous genes encoding iron transport proteins. The genes *yfeABCD*, *fbpABC*, *fecBCD*, *tonB*, and *exbBD* all showed increased levels of expression of between 2.1- and 7.5-fold under conditions of iron limitation. In addition, the genes encoding the global stress response proteins RecX, HslV and HtrA showed increased expression under iron limitation, as did genes encoding the cell surface/membrane biogenesis proteins (GmhA, Skp, HexC, and PonB). Together, these data suggest that growth in the presence of reduced iron results in structural effects on the bacterial outer membrane. Interestingly, 27% of the genes showing altered expression encoded proteins with no characterised function (Paustian et al. 2001). Today, ten years after these initial studies were completed, the majority of these proteins remains uncharacterised. However, current analysis of the 27 hypothetical proteins identified as up-regulated showed that ten have domains now known or predicted to be associated with iron binding and/or transport.

The gene expression pattern of *P. multocida* in response to the sole iron sources haemoglobin, transferrin, ferric citrate and ferritin was also assessed by DNA microarrays (Paustian et al. 2002). Growth in the presence of transferrin had the most significant global effect on *P. multocida* gene transcription, with 10% of genes showing increased expression and 8% showing decreased expression, while growth in the presence of ferritin resulted in increased or reduced expression of only 5% of genes. Surprisingly, only two genes, namely *ptsN* and *sapD*, showed altered expression under all of the tested conditions and many known iron acquisition genes did not show altered transcription under any of the conditions. The ferric uptake regulator Fur, was expressed at elevated levels in all conditions except when haemoglobin was used as the sole iron source.

While the coordinated transcriptional response of *P. multocida* to either low iron or defined iron sources is likely to be important for both bacterial survival in the host and disease pathogenesis, the true response to the host environment can only be assessed using animal studies. To this end, the Pm70 DNA microarrays have also been used to characterise the bacterial response to growth in host tissues; either in the blood or livers of infected chickens (Boyce et al. 2002, 2004). As *P. multocida* reaches very high numbers in both tissues late in infection ($>1 \times 10^9$ CFU/gram of tissue), it is possible to purify sufficient bacterial RNA for DNA microarray experiments. Indeed, as a consequence of the ability to recover sufficient RNA from these in vivo samples, the *P. multocida* experiments were among the first to characterise the transcriptional response of a pathogen while growing in its natural host. For these experiments, the virulent fowl cholera-causing strain X73 was used to infect chickens and bacteria recovered from either the blood or liver tissue of birds in the late stages of disease. The *P. multocida* transcriptional response under these in vivo conditions was then compared with the transcriptome of bacteria grown in vitro in rich medium (brain heart infusion) at the same temperature (41°C, the normal body temperature of chickens).

Analysis of the gene expression pattern of *P. multocida* recovered from the blood of three infected chickens, indicated that 17 genes were increased in expression and 23 genes were decreased in expression during growth in blood as compared to growth in vitro in rich medium (Boyce et al. 2002). A number of the up-regulated genes (*gdhA*, *asnA*, *aspC*, *ilvH, dppA*) encoded proteins predicted to be involved in the biosynthesis, transport and conversion of amino acids. Furthermore, the nitrate reductase (*napF*, *napA*, *napC* and *napB*) operon was also strongly expressed in vivo; this pathway may be used for nitrate scavenging and the biosynthesis of glutamate. These data strongly suggest that there is a limited supply of amino acids in the blood (at least during the late stages of infection) and that amino acid synthesis is critical for growth in this niche. The gene expression profile of *P. multocida* recovered from infected liver tissue was in general statistically similar to that observed in bacteria recovered from blood, although a significantly larger set (49 genes) was up-regulated during growth in liver (Boyce et al. 2002). There was a significant increase in expression of genes encoding proteins involved in anaerobic metabolism, suggesting that the liver environment is more anaerobic than the in vitro control growth conditions, and an increase in expression of a number of genes involved in carbohydrate transport and metabolism, suggesting a shift to different carbohydrate resources for energy production. Perhaps, unsurprisingly, most of the genes identified by the in vivo microarray analyses were involved in metabolic processes required directly for *P. multocida* survival in the in vivo niche. While it is possible that some of the uncharacterised proteins identified by these analyses may encode true virulence factors, experiments to confirm this have not been reported. Indeed, recovering bacteria from earlier in infection may be more useful for the identification of virulence factors; unfortunately, experimental limitations mean that whole-genome analyses at very early time points are not possible at present.

Whole-genome DNA microarrays have also been used to analyse the response of *P. multocida* to the presence of antibiotics (Melnikow et al. 2008; Nanduri et al. 2009). The transcriptional response was measured following exposure to novobiocin, tilmicosin, florfenicol, rifampin, trimethoprim, brodimoprim, cefquinome, enrofloxacin or thiazine (all at 1 × the minimal inhibitory concentration (MIC)) and with enrofloxacin, amoxycillin and chlortetracycline (at ¼ × MIC). When the bacteria were exposed to each antibiotic at 1 × MIC, there was a major difference in the number of differentially regulated genes observed in response to the type of antibiotic used (Melnikow et al. 2008); bacteriostatic antibiotics resulted in large numbers of differentially expressed genes, whereas bactericidal antibiotics altered the regulation of very few genes. Indeed, addition of novobiocin resulted in altered expression of 39% of the genome whereas thiazine, enrofloxacin or cefquinome resulted in altered expression of less than 1.5% of the genome. Interestingly, the expression of many known virulence genes was reduced in the presence of antibiotics. Capsule biosynthesis genes showed reduced expression following all 1 × MIC antibiotic treatments except when rifampin was used. Similarly, the iron transport genes *exbB*, *exbD* and *tonB* also showed reduced expression under almost all tested conditions. These data suggest that even without complete killing of a bacterial population many antibiotics may have effects through the suppression of *P. multocida* virulence gene expression.

7 Proteomic Analyses

The determination of the Pm70 genome sequence was also a prerequisite for complete proteomic studies, as the identification of proteins by mass spectrometry (peptide mass fingerprinting) requires knowledge of the sequence of all the proteins expressed by the strain. Initial proteomic studies of *P. multocida* characterised the protein components of the *P. multocida* outer membrane and determined how expression of these outer membrane proteins was altered under either low iron conditions or growth in the blood of infected chickens (Boyce et al. 2006). These analyses identified more than 30 proteins likely to be legitimate components of the *P. multocida* outer membrane, with the major outer membrane protein identified as Pm0786. Eleven of the outer membrane proteins were predicted lipoproteins, ten were predicted porins and half of these were predicted to be involved in iron uptake. One of these proteins, Pm0803, was only identified in the outer membrane fractions of bacteria grown under low iron conditions or recovered from the blood of infected chickens, indicating that it might be critical for in vivo survival. The gene encoding this protein was also identified as strongly up-regulated under low iron and in vivo growth conditions using DNA microarrays (Boyce et al. 2004; Paustian et al. 2001). However, a *pm0803* mutant is not attenuated for virulence (our unpublished data), consistent with functional redundancy in *P. multocida* iron transport systems.

Proteomic methods have also been applied to the analysis of the *P. multocida* response to sub-inhibitory (1/4 × MIC) concentrations of antibiotic (Nanduri et al. 2006, 2008). Following antibiotic addition, *P. multocida* proteins were identified by two-dimensional liquid chromatography-electrospray ionisation tandem mass spectrometry (2D-LC-ESI-MS/MS) using either unlabelled peptides, or peptides labelled with isotope-coded affinity tags (ICAT). Experiments using unlabelled peptides followed by 2D-LC-ESI-MS/MS identified more than 50% of the *P. multocida* Pm70 proteins (Nanduri et al. 2008) while the similar analysis using ICAT labelled peptides identified only 20% of the possible proteins (Nanduri et al. 2006). The more complete analysis (Nanduri et al. 2008) identified 147, 126 and 134 proteins as differentially expressed following treatment with amoxicillin, chlortetracycline or enroflaxacin, respectively. A strong shock response was detected with an increase in expression of the proteins GroES (following all treatments) and GroEL and DnaK (following amoxicillin treatment). Although a decrease in production of a small number of known virulence factors, including OmpA, was observed, the number of differentially expressed virulence factors was much less than was observed in similar transcriptional profiling experiments (see above).

8 Concluding Remarks

The publication and release of the first complete *P. multocida* genome in 2001 facilitated new approaches for studying *P. multocida* pathogenesis. Genomic analyses immediately identified the very large genetic repertoire of the organism for iron acquisition and identified the significant proportion of the genome used for production of the virulence factors, filamentous haemagglutinin and capsule. Furthermore, the use of whole-genome microarrays facilitated the dissection of how *P. multocida* adapts to survive under conditions of low free iron and within certain in vivo niches. However, the generation of multiple *P. multocida* genome sequences has lagged behind many other Gram-negative pathogens. Therefore, thorough comparative genomic analyses allowing for the identification of core and accessory *P. multocida* genes and/or disease-specific factors have not been possible. Only recently have new *P. multocida* genomes become available and it is hoped that once these are closed we can begin to define the genes responsible for *P. multocida* host specificity and identify those genes which make *P. multocida* the exquisitely adapted pathogen that it is.

References

Ahir VB, Roy A, Jhala MK, Bhanderi BB, Mathakiya RA, Bhatt VD, Padiya KB, Jakhesara SJ, Koringa PG, Joshi CG (2011) Genome sequence of *Pasteurella multocida* subsp. *gallicida* Anand1_poultry. J Bacteriol 193:5604

Alikhan NF, Petty NK, Ben Zakour NL, Beatson SA (2011) BLAST ring image generator (BRIG): simple prokaryote genome comparisons. BMC Genomics 12:402

Boyce JD, Cullen PA, Nguyen C, Wilkie I, Adler B (2006) Analysis of the *Pasteurella multocida* outer membrane sub-proteome and its response to the in vivo environment of the natural host. Proteomics 6:870–880

Boyce JD, Wilkie I, Harper M, Paustian ML, Kapur V, Adler B (2002) Genomic scale analysis of *Pasteurella multocida* gene expression during growth within the natural chicken host. Infect Immun 70:6871–6879

Boyce JD, Wilkie I, Harper M, Paustian ML, Kapur V, Adler B (2004) Genomic-scale analysis of *Pasteurella multocida* gene expression during growth within liver tissue of chickens with fowl cholera. Microbes Infect 6:290–298

Campoy S, Aranda J, Alvarez G, Barbe J, Llagostera M (2006) Isolation and sequencing of a temperate transducing phage for *Pasteurella multocida*. Appl Environ Microbiol 72: 3154–3160

Darling AE, Mau B, Perna NT (2010) progressiveMauve: multiple genome alignment with gene gain, loss and rearrangement. PLoS One 5:e11147

Huson DH, Bryant D (2006) Application of phylogenetic networks in evolutionary studies. Mol Biol Evol 23:254–267

iDali C, Foged NT, Frandsen PL, Nielsen MH, Elling F (1991) Ultrastructural localization of the *Pasteurella multocida* toxin in a toxin-producing strain. J Gen Microbiol 137:1067–1071

Kehrenberg C, Schwarz S (2005) Plasmid-borne florfenicol resistance in *Pasteurella multocida*. J Antimicrob Chemother 55:773–775

Kehrenberg C, Tham NT, Schwarz S (2003) New plasmid-borne antibiotic resistance gene cluster in *Pasteurella multocida*. Antimicrob Agents Chemother 47:2978–2980

May BJ, Zhang Q, Li LL, Paustian ML, Whittam TS, Kapur V (2001) Complete genomic sequence of *Pasteurella multocida*, Pm70. Proc Natl Acad Sci U S A 98:3460–3465

Melnikow E, Schoenfeld C, Spehr V, Warrass R, Gunkel N, Duszenko M, Selzer PM, Ullrich HJ (2008) A compendium of antibiotic-induced transcription profiles reveals broad regulation of *Pasteurella multocida* virulence genes. Vet Microbiol 131:277–292

Michael GB, Kadlec K, Sweeney MT, Brzuszkiewicz E, Liesegang H, Daniel R, Murray RW, Watts JL, Schwarz S (2012a) ICE*Pmu1*, an integrative conjugative element (ICE) of *Pasteurella multocida*: analysis of the regions that comprise 12 antimicrobial resistance genes. J Antimicrob Chemother 67:84–90

Michael GB, Kadlec K, Sweeney MT, Brzuszkiewicz E, Liesegang H, Daniel R, Murray RW, Watts JL, Schwarz S (2012b) ICE*Pmu1*, an integrative conjugative element (ICE) of *Pasteurella multocida*: structure and transfer. J Antimicrob Chemother 67:91–100

Nanduri B, Lawrence ML, Boyle CR, Ramkumar M, Burgess SC (2006) Effects of subminimum inhibitory concentrations of antibiotics on the *Pasteurella multocida* proteome. J Proteome Res 5:572–580

Nanduri B, Lawrence ML, Peddinti DS, Burgess SC (2008) Effects of subminimum inhibitory concentrations of antibiotics on the *Pasteurella multocida* proteome: a systems approach. Comp Funct Genomics, Article ID 254836

Nanduri B, Shack LA, Burgess SC, Lawrence ML (2009) The transcriptional response of *Pasteurella multocida* to three classes of antibiotics. BMC Genomics 10(Suppl 2):S4

Nielsen JP, Rosdahl VT (1990) Development and epidemiological applications of a bacteriophage typing system for typing *Pasteurella multocida*. J Clin Microbiol 28:103–107

Paustian ML, May BJ, Cao D, Boley D, Kapur V (2002) Transcriptional response of *Pasteurella multocida* to defined iron sources. J Bacteriol 184:6714–6720

Paustian ML, May BJ, Kapur V (2001) *Pasteurella multocida* gene expression in response to iron limitation. Infect Immun 69:4109–4115

Pullinger GD, Bevir T, Lax AJ (2004) The *Pasteurella multocida* toxin is encoded within a lysogenic bacteriophage. Mol Microbiol 51:255–269

San Millan A, Escudero JA, Gutierrez B, Hidalgo L, Garcia N, Llagostera M, Dominguez L, Gonzalez-Zorn B (2009) Multiresistance in *Pasteurella multocida* is mediated by coexistence of small plasmids. Antimicrob Agents Chemother 53:3399–3404

Subaaharan S, Blackall LL, Blackall PJ (2010) Development of a multi-locus sequence typing scheme for avian isolates of *Pasteurella multocida*. Vet Microbiol 141:354–361

Tatum FM, Yersin AG, Briggs RE (2005) Construction and virulence of a *Pasteurella multocida* *fhaB2* mutant in turkeys. Microb Pathog 39:9–17

Townsend KT, Boyce JD, Chung JY, Frost AJ, Adler B (2001) Genetic organization of *Pasteurella multocida cap* loci and the development of a multiplex capsular PCR typing system. J Clin Microbiol 39:924–929. In: Erratum. J Clin Microbiol 939:2378
Wright CL, Strugnell RA, Hodgson AL (1997) Characterization of a *Pasteurella multocida* plasmid and its use to express recombinant proteins in *P. multocida*. Plasmid 37:65–79

The Key Surface Components of *Pasteurella multocida*: Capsule and Lipopolysaccharide

Marina Harper, John D. Boyce and Ben Adler

Abstract The capsule and lipopolysaccharide (LPS) of *Pasteurella multocida* constitute the major components of the bacterial cell surface. As well as forming the basis for the most widely used classification systems, they play key roles in a range of interactions between the bacteria and the hosts they colonize or infect. Both polysaccharides are involved in the avoidance of host innate immune mechanisms, such as resistance to phagocytosis, complement-mediated killing, and the bactericidal activity of antimicrobial peptides; they are therefore essential for virulence. In addition, LPS is a major antigen in the stimulation of adaptive immune responses to infection.

Contents

M. Harper (✉) · J. D. Boyce · B. Adler
Australian Research Council Centre of Excellence in Structural and Functional Microbial Genomics, Department of Microbiology, Monash University, Clayton, Victoria 3800, Australia
e-mail: ben.adler@monash.edu

Current Topics in Microbiology and Immunology (2012) 361: 39–51
DOI: 10.1007/82_2012_202

Published Online: 29 February 2012

1 The Capsule of *Pasteurella multocida*

Pasteurella multocida has long been differentiated into five distinct capsular types (Carter 1967; Rimler and Rhoades 1987), designated A, B, D, E, and F. There is a well recognized and documented association of capsule type with particular hosts and diseases. Fowl cholera is most commonly associated with type A strains, while hemorrhagic septicemia is caused only by types B and E. Atrophic rhinitis strains belong to type D. Intriguingly, the molecular and cellular basis for these host and disease associations remains unknown. Nevertheless, much progress has been made in recent years in elucidating the structures of some capsule types, the genetic basis for capsule biosynthesis and the role of capsules in pathogenesis.

2 Chemical Structure of *Pasteurella multocida* Capsules

All *P. multocida* capsules are carbohydrate polymers. The capsule of type A strains has long been known to be composed of hyaluronic acid (HA; Carter 1958), which is indistinguishable from the mammalian or avian HA, a high molecular mass repeating polymer of glucuronic acid and N-acetyl glucosamine, that forms a major component of the extracellular matrix. It is thus difficult to imagine that an antibody response against HA could be elicited. Given that strains were identified as type A based on reactivity with so-called capsule-specific antisera, it is unlikely that the reactive material in the very crude preparations used was indeed HA.

The capsules of types D and F were known to be composed of polymers that were related to, but distinct from, the type A HA. These conclusions were based on their susceptibility profiles to hydrolysis by specific enzymes such as heparinase, hyaluronidase, and chondroitinase (Pandit and Smith 1993; Rimler 1994). Some years elapsed before their definitive identification as heparin and chondroitin, respectively (DeAngelis et al. 2002). Both capsules are linear polysaccharides made up of repeating disaccharide units. The type F polymer consists of N-acetyl galactosamine and glucuronic acid while the type D heparin contains the same disaccharide unit as HA, but with a different inter-monosaccharide linkage.

The capsules of types B and E remain the least well characterized. Compositional analyses of type B have reported the presence of mannose, arabinose, and galactose (Muniandy et al. 1992). However, the method of preparation used in that study would not detect any amino or other modified sugars. Indeed, the genetic locus for type B capsule biosynthesis suggests the presence of N-acetyl mannosaminuronic acid rather than mannose (Boyce et al. 2000). There have been no reported analyses of type E capsule.

3 Genetics of Capsule Biosynthesis

The genetic organization of the biosynthetic loci for all five capsule types has been determined (Boyce et al. 2000, 2010; Chung et al. 1998; Townsend et al. 2001), revealing an arrangement similar to that for typical Group II capsule biosynthetic regions found in other Gram-negative bacteria. All loci contain a region, designated Region 1, which comprises four highly conserved genes (designated *hexABCD* or *cexABCD*; Boyce et al. 2010) which encode the components of an ABC transport system required for export and surface presentation of the capsule polymer. Inactivation of *hexA* or *cexA* abrogated capsular transport and resulted in the intracellular accumulation of capsular polysaccharide (Boyce and Adler 2000; Chung et al. 2001). Similarly, the two-gene locus (*phyAB* or *lipAB*), Region 3, which encodes proteins required for anchoring of the capsule polysaccharide to the bacterial cell surface, is also highly conserved across all five types, although in the type B and E loci *lipA* is located adjacent to Region 1 (see Boyce et al. 2010 for diagrammatical representations of capsule biosynthesis loci).

In contrast, the intervening Region 2 contains genes specific for each capsule serogroup. The encoded proteins catalyze the synthesis and assembly of the corresponding polysaccharide; their functions have been confirmed experimentally for capsule types A, B, D, and F (Boyce and Adler 2000; DeAngelis et al. 1998, 2002; Kane et al. 2006). Interestingly, the transferases involved in the assembly of HA, heparin, and chondroitin (types A, D and F) are all bifunctional enzymes which catalyze the addition of both the monosaccharide constituents to the polysaccharide polymer.

Subsequently a second, cryptic heparin synthase gene was identified outside the capsule biosynthetic locus in a type D strain and shown to encode a functional heparin synthase (DeAngelis and White 2004). Homologous genes were identified in most type A, D, and F isolates. It was suggested that these multiple polysaccharide synthases may facilitate variation in capsule composition, but neither this possibility nor the role of these additional genes in virulence has been investigated experimentally.

The elucidation of the genetic basis for capsule biosynthesis has facilitated the development of a much improved method for laboratory typing of *P. multocida* isolates utilizing a multiplex PCR assay based on specific gene sequences within Region 2 (Townsend et al. 2001). The PCR-based typing method has therefore supplanted the previously unreliable conventional serotyping (Arumugam et al. 2011) and is now used extensively worldwide as the gold standard (Dziva et al. 2008; Jaglic et al. 2004; Varga et al. 2007).

4 Regulation of Capsule Production

Differences in the amount of capsule produced by *P. multocida* under a range of nutritional and environmental conditions have long been recognized. For example, Jacques et al. (1994) reported reduced production of capsule under low iron concentration in vitro. However, the mechanism(s) responsible for regulating capsule biosynthesis remained enigmatic until the very recent discovery that the global transcriptional regulator, Fis, was essential for the transcription of all regions of the type A capsule biosynthetic locus (Steen et al. 2010). Fis mutants produced no capsule and were completely attenuated for virulence in mice and chickens. Fis was shown to act on both of the promoters known to drive transcription of Region 1 and Regions 2/3. Interestingly, Fis was shown to also activate the transcription of at least 30 additional genes (20 operons), including some encoding known virulence factors such as filamentous hemagglutinin (FhaB2) or the protective membrane protein PlpE (Wu et al. 2007).

5 The Role of Capsule in Virulence

The capsule of *P. multocida* is essential for virulence. It is difficult to draw definitive conclusions from early work with spontaneously derived acapsular mutants because isogenic pairs of strains were not available and the nature of the genetic change(s) could not be defined. However, genetically defined mutants which are unable to either synthesize or export capsule to their surface are attenuated for virulence (Boyce and Adler 2000; Chung et al. 2001). Intriguingly, the specific role of capsule may vary between different capsule types. For example, a type A acapsular mutant became highly sensitive to complement-mediated killing by chicken serum, whereas no difference was observed in susceptibility to mouse or cattle serum with a type B acapsular mutant. However, the type B mutant was significantly more sensitive to phagocytosis by mouse macrophages in vitro.

There are conflicting reports in the literature on the possible role of capsule in adhesion to host cells and tissues. For example, there are studies which argue against a role in adhesion (Glorioso et al. 1982), while others advocate such a role (Pruimboom et al. 1996, 1999). However, these studies did not use isogenic, genetically defined strains. See "*Pasteurella moltocida*: Diseases and Pathogenesis" for details. Given the similarity of type A, D, and F capsules to normal host tissue polymers, it is difficult to envision a role for capsule in adhesion. Indeed, in the related pathogen *Haemophilus influenzae*, capsule expression is downregulated at the mucosal surface to allow presentation of adhesins and then upregulated upon entry into the bloodstream to mediate resistance to complement and phagocytosis (St Geme 2002). There have been no equivalent studies for *P. multocida*.

6 The Lipopolysaccharide of *Pasteurella multocida*

Pasteurella multocida is currently classified into 16 Heddleston serovars (formerly serotypes) based on Lipopolysaccharide (LPS) antigen (Heddleston and Rebers 1972). LPS forms the major component of the outer leaflet of the Gram-negative outer membrane, where the tight packing of LPS molecules forms an impermeable hydrophobic barrier. LPS can be divided into three distinct regions, the hydrophobic lipid A domain, a non-repeating inner and outer core oligosaccharide, and an O-antigen, consisting of a repeating oligosaccharide (Raetz and Whitfield 2002). *P. multocida* LPS does not contain an O-antigen and in that respect is similar to the LPS or lipooligosaccharide (LOS) expressed by a number of other important pathogens, including *Campylobacter jejuni*, *Neisseria* spp., *Haemophilus influenzae,* and *Moraxella catarrhalis*. As there is no O-antigen produced by these bacteria, the most distal and variable component of the LPS is the outer core region. Indeed, nearly all of the Heddleston type strains express LPS molecules with different outer core structures. However, the inner core structure is highly conserved across *P. multocida* strains.

7 The LPS Inner Core Structure

The LPS structures produced by the 16 Heddleston type strains have been determined (Harper et al. 2011a, b; St Michael et al. 2005a–c) (A. Cox unpublished data). In all strains examined, the inner core of the LPS molecule is highly conserved, but many strains simultaneously express two different LPS forms, termed glycoform A and glycoform B that share an identical outer core structure but differ in the structure of the inner core (St Michael et al. 2005b). The inner core structure of glycoform A contains a single phosphorylated 3-deoxy-D-mannooctulosonate (Kdo) residue attached to the lipid A, whereas the inner core of glycoform B contains two Kdo residues (Fig. 1). Attached to the first Kdo residue in the inner core of both glycoforms is a tri-heptose side chain and once this is assembled a glucose residue is attached to the four position of the first heptose to form the main branch of the LPS structure. Glycoform A also contains a second glucose attached to the six position of the first heptose (Fig. 1; St Michael et al. 2005b). The glycoform A inner core is conserved amongst all *P. multocida* strains and is also shared by two other animal-associated pathogens within the Pasteurellaceae family, *Mannheimia haemolytica* and *Actinobacillus pleuropneumoniae* (St Michael et al. 2011).

8 Genetics and Assembly of the LPS Inner Core

Bioinformatic analysis of available *P. multocida* genomes, in combination with LPS structural analysis of defined mutants, has allowed the identification of all the transferases required for the assembly of the inner core in the serovar 1 strain, VP161 (Boyce et al. 2009; Harper et al. 2007a, b). The Kdo transferase KdtA, which adds Kdo residues to lipid A, was identified based on a high degree of

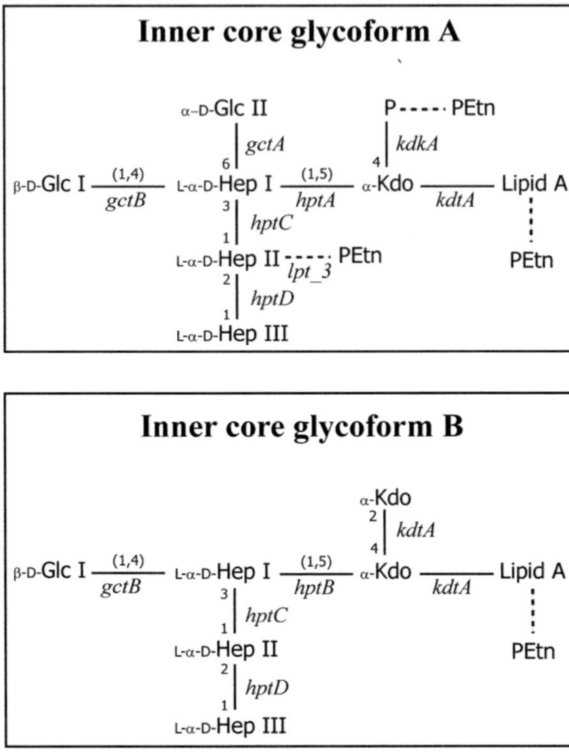

Fig. 1 Schematic representation of the two *P. multocida* LPS inner core structures. The gene encoding the transferase required for each step in the inner core assembly is shown beside/underneath each linkage. Two LPS glycoforms, termed glycoform A and glycoform B, are produced by most *P. multocida* strains examined. The inner core of glycoform B contains lipid A and two 3-deoxy-D-mannooctulosonate (Kdo) residues, whereas the inner core of glycoform A consists of a lipid A and a single phosphorylated Kdo residue. Glycoform A also has an additional glucose residue attached to the six position of the first heptose. The specific number and position of PEtn residues attached to each LPS structure is serovar, strain and glycoform-type dependent. Residues are KDO, 3-deoxy-D-mannooctulosonate; P, phosphate; PEtn, phosphoethanolamine; Glc, glucose; Hep, heptose. Reproduced with permission from Elsevier B.V, Veterinary Microbiology (2011), 153: pp 109–115

identity to the dual Kdo transferase in *E. coli* (Belunis and Raetz 1992). The phosphorylation of the first Kdo residue is performed by the kinase, KdkA, and this addition dictates glycoform assembly; once the glycoform A acceptor molecule (Lipid A-Kdo-P) is generated, the addition of the second Kdo residue can no longer occur. However, if KdtA adds a second Kdo residue before phosphorylation of the first, then a glycoform B acceptor molecule (Lipid A-Kdo_2) is generated (Harper et al. 2011a). Experiments with in vivo-derived *P. multocida kdkA* mutants have identified residues within KdkA that are essential for kinase activity (Harper et al. 2010).

As there are two structurally distinct acceptor molecules, two transferases are required for the addition of the first heptose. The heptosyltransferase, HptA, donates heptose only to the glycoform A acceptor molecule, whereas HptB is specific for the glycoform B acceptor molecule (Harper et al. 2007a). Once the first heptose is attached to the nascent LPS molecules the second and third heptose sugars are added to the side chain of both glycoforms by the transferases HptC and HptD, respectively. A complete heptose side chain is critical for the full activity of GctB, the next transferase involved in assembly which donates glucose to the main chain of the LPS molecule (Harper et al. 2004). Finally, the addition of PEtn to the second heptose produced by some strains of *P. multocida* is performed by the transferase Lpt_3. The addition of this PEtn is significant. Experiments have shown that monoclonal antibodies raised against serovar 2 LPS (lacking the heptose II-linked PEtn) are unable to bind to the serovar 5 LPS, identical except for a heptose II-linked PEtn, clearly indicating that the addition of PEtn to the inner core of the serovar 5 LPS is sufficient for the LPS to be immunologically distinct (St Michael et al. 2009).

Analysis of the *P. multocida* genome reveals that the genes required for the assembly of the inner core are located in several regions of the genome; KdtA, KdkA, HptA, and GctB are located within the one locus, while the genes for HptB and HptC are encoded together elsewhere on the genome (Harper et al. 2007a). The three remaining transferase genes required for the assembly of the inner core (encoding HptD, GctA and Lpt_3) are each located separately on the *P. multocida* genome (Harper et al. 2007a; St Michael et al. 2009).

9 The LPS Outer Core

Genetic analysis of *P. multocida* strains reveals that the genes required for the biosynthesis and assembly of the outer core region (defined as the LPS structure beyond the first glucose) of the LPS are located between the conserved non-LPS related genes *priA* and *fpg* (Boyce et al. 2009; Harper et al. 2011c; St Michael et al. 2005c, 2009). The LPS outer core structures produced by *P. multocida* vary considerably. The serovar 1 strains, X73 and VP161, elaborate a relatively short outer core consisting of a heptose substituted at the four and six positions with galactose. Phosphocholine is added to the three position of each galactose, and in strain X73 only, two PEtn residues are attached to each terminal galactose (Fig. 2; St Michael et al. 2005a). The dual transferase required for the addition of these PEtn residues is encoded elsewhere on the genome (our unpublished data). All of the transferases required for the assembly of the outer core in the serovar 1 strain VP161 have been identified and characterized (Boyce et al. 2009; Harper et al. 2007b). These include HptE, that adds heptose to the inner core glucose; GatA, a predicted dual transferase that transfers both galactose residues to the heptose and a unique phosphocholine transferase, PcgD. Genetically related to the serovar 1 strains is the serovar 14 type strain P2225, which shares the same outer

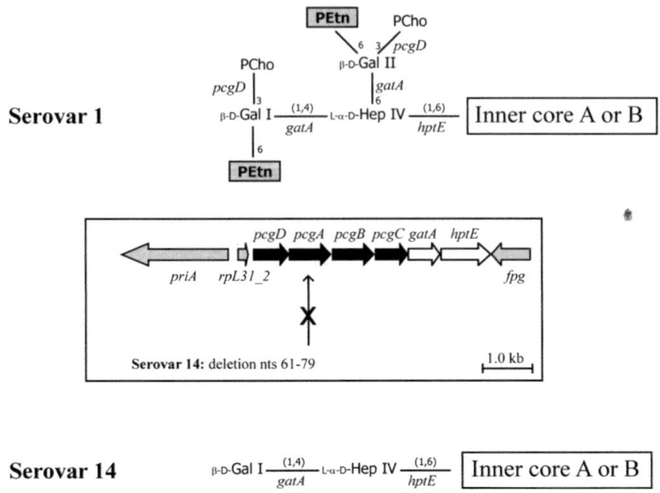

Fig. 2 Schematic representation of the LPS outer core oligosaccharide structure expressed by *P. multocida* strains belonging to serovar 1 (X73 and VP161) and 14 (P2225). The conserved inner core of the LPS molecule is represented by a *box*. The gene encoding the glycosyltransferase required for each step in LPS synthesis is shown beside/underneath each linkage. The phosphoethanolamine (PEtn) residues (*boxed* and *shaded gray*) attached to Gal I and Gal II are observed only in LPS derived from the serovar 1 type strain X73. Residues are Hep, heptose; Gal, galactose, PCho, phosphocholine, PEtn, phosphoethanolamine. The *inset* shows the genetic organization of the serovar 1 and 14 LPS outer core biosynthesis loci. Conserved genes unrelated to LPS biosynthesis are shown in *gray*. LPS outer core biosynthesis genes are shown in *black* (PCho biosynthesis operon) or *white* (glycosyltransferase genes). Reproduced with permission from Elsevier B.V, Veterinary Microbiology (2011), 153: pp 109–115

core biosynthesis locus, but contains a 22 base pair deletion within the phosphocholine biosynthesis gene, *pcgA* (Harper et al. 2011b). As a result of this mutation, P2225 is unable to add phosphocholine to its LPS. The loss of phosphocholine has a knock-on effect on the activity of the dual galactosyltransferase GatA, as an earlier study using a serovar 1 phosphocholine mutant revealed that there was also a loss of galactose at the six position of the fourth heptose in the absence of PCho addition (Harper et al. 2007b). As a result, despite the two strains sharing the same outer core biosynthesis locus, the mutation within *pcgA* in the serovar 14 locus resulted in a significant truncation of the LPS structure (Fig. 2).

The type strains of serovars 2 and 5 share an identical LPS outer core consisting of αHep-(1-6)-αHep-(1-6)-βGlcNAc-(1-4)-βGal. However, these serovars can be antigenically differentiated as a result of the non-stoichiometric addition of PEtn to the second heptose of the inner core of the serovar 5 LPS (performed by the transferase Lpt_3) (St Michael et al. 2009).

The outer core of the LPS produced by the serovar 3 strain Pm70 terminates with a structure identical to the Forssman antigen expressed on the surface of mammalian cells, consisting of βGalNAc-(1-3)-βGalNAc-(1-3)-αGal-(1-4)-βGal. There is also strong evidence that the LPS expressed by the serovar 3 strains,

Pm70 and P1059, are decorated with sialic acid (Steenbergen et al. 2005; Tatum et al. 2009). In our own work, genetic analyses has revealed that the type strains of serovars 3 and 4 share the same outer core biosynthesis locus as Pm70 but as a result of a mutation within this locus, serovar 4 LPS is significantly truncated. Interestingly, there is no obvious mutation within the locus encoded by serovar 3 type strain but it does not elaborate the terminal residue present on the Pm70 structure (A. Cox; our unpublished data).

One of the largest *P. multocida* LPS structures reported to date is expressed by the serovar 9 type strain P2095, originally isolated from a turkey. The outer core of serovar 9 consists of heptose, rhamnose, and 3-acetamido-3,6-dideoxy-α-D-glucose (Qui3NAc), a deoxyamino sugar rarely reported to be a component of LPS. The LPS outer core biosynthesis locus contains all the genes necessary for both rhamnose and Qui3NAc biosynthesis including a predicted bifunctional enzyme (QdtD) that has both dTDP-4-oxo-6-deoxy-D-glucose-3,4-oxoisomerase and dTDP-3-oxo-6-deoxy-α-D-glucose transacetylase activity (Harper et al. 2011c).

Of the nine remaining Heddleston serovar type strains, four express full length LPS outer cores and five express truncated versions of these "parent" structures (A. Cox, unpublished data). Analysis of the outer core biosynthesis loci indicated that the assembly of truncated LPS structures in each case was due to a mutation within an LPS-associated gene. Indeed, only eight different genetic loci encode all of the LPS structures expressed by the 16 Heddleston serovars (our unpublished data).

10 The Role of LPS in Virulence

Pasteurella multocida LPS has a clearly defined role in disease. This was first shown when a *galE* mutant lacking the ability to synthesize UDP-galactose, essential for *P. multocida* LPS biosynthesis, was completely attenuated in mice (Fernandez de Henestrosa et al. 1997). Following this study, two independent signature-tagged mutagenesis banks constructed in strains belonging to serovars 1 and 3 were screened for attenuation in both mice and chickens (serovar 1) or mice only (serovar 3) (Fuller et al. 2000; Harper et al. 2003). In each study, a fully attenuated mutant containing a transposon insertion in an LPS glycosyltransferase gene was identified. Final proof that LPS was an essential virulence factor in fowl cholera was demonstrated using directed mutants of the highly virulent serovar 1 strain VP161. VP161 LPS mutants expressing severely truncated LPS lacking all of the outer core structure were completely attenuated in the natural chicken host (Harper et al. 2007a, 2004). Furthermore, a complete outer core structure is also essential for strain VP161 to exhibit full virulence; a *pcgC* mutant lacking the distal end of the LPS molecule (both PCho residues and one galactose residue) was significantly attenuated in chickens and was more susceptible to the chicken antimicrobial peptide fowlicidin 1 (Harper et al. 2007b).

Although significantly attenuated with respect to systemic infection, most LPS outer core mutants were able to persist for days in muscle tissue at the site of injection (Boyce et al. 2009). In contrast, LPS inner core mutants are cleared rapidly from the injection site. Interestingly, an outer core LPS mutant harboring an *hptE* mutation produced an alternative glycoform, termed glycoform C, that contains the outer core oligosaccharide β-Gal-(1-4)-β-GlcNAc-(1-3)-β-Gal-(1-3)-β-GlcNAc attached to a glycoform A inner core (Boyce et al. 2009). The expression of this alternative glycoform on the surface of the *hptE* mutant may contribute to in vivo survival, as the *hptE* mutant was the only outer core mutant present in significant numbers in the blood of infected chickens (Boyce et al. 2009). Low levels of glycoform C have also been detected in wild-type strains of *P. multocida* belonging to number of different serovars (A. Cox, unpublished data) and may be an example of molecular mimicry as it contains β-Gal-(1-4)-β-GlcNAc-(1-3), a key structural component of eukaryotic glycophosholipids.

11 The Role of LPS in Immunity

Pasteurella multocida infects a wide range of animals and the infection may present with a diverse range of clinical signs. Most of the investigations into LPS and immunity have used fowl cholera strains in poultry. Accordingly, this section will focus on the role of LPS in immunity to fowl cholera in birds.

LPS has a direct role in defending the organism against the avian innate immune system. Loss of even the outermost residues (phosphocholine) is sufficient to increase the susceptibility of the strain to the cathelicidin antimicrobial, fowlicidin 1 (Harper et al. 2007b). However, the most dramatic loss in resistance was observed in LPS mutants that had lost part or all of the inner core heptose side chain (Harper et al. 2007a). The Gram-negative bacterial component initially targeted by cathelicidin antimicrobials is believed to be LPS (Turner et al. 1998) and likely to be the negatively charged lipid A component. It is predicted that the increased susceptibility to fowlicidin 1 exhibited by *P. multocida* LPS mutants is due to an increase in negative charge and/or major conformational changes in the LPS structure leading to exposure of lipid A. Unlike the capsule of *P. multocida*, LPS does not play an obvious role in preventing complement-mediated killing; early studies using a serovar 1 *hptD* mutant expressing a severely truncated LPS demonstrated that this strain was able to grow as well as the wild-type parent in fresh chicken serum (Harper et al. 2004).

Acquired immunity to *P. multocida* is mainly antibody mediated, although there is some evidence that cell-mediated immunity in chickens may also play a role (Baba 1984). However, this work has not been reproduced. Vaccination with bacterins induces LPS-specific immunity which generally protects only against challenge with strains belonging to the same serovar (Boyce et al. 2010). Antibodies against LPS are protective, but are also serovar-specific; initial bacterin vaccination experiments in our laboratory using mutants expressing different truncated LPS molecules revealed that the immune responses elicited by these LPS mutants were not protective against challenge with an isogenic strain expressing the full-length LPS molecule (our unpublished data).

Acknowledgments Original work in the authors' laboratory was supported by the Australian Research Council.

References

Arumugam ND, Ajam N, Blackall PJ, Asiah NM, Ramlan M, Maria J, Yuslan S, Thong KL (2011) Capsular serotyping of *Pasteurella multocida* from various animal hosts—a comparison of phenotypic and genotypic methods. Trop Biomed 28:55–63

Baba T (1984) Cell-mediated immune protection in chickens against *Pasteurella multocida*. Res Vet Sci 36:225–230

Belunis CJ, Raetz CR (1992) Biosynthesis of endotoxins. Purification and catalytic properties of 3-deoxy-D-manno-octulosonic acid transferase from *Escherichia coli*. J Biol Chem 267: 9988–9997

Boyce JD, Adler B (2000) The capsule is a virulence determinant in the pathogenesis of *Pasteurella multocida* M1404 (B:2). Infect Immun 68:3463–3468

Boyce JD, Chung JY, Adler B (2000) Genetic organisation of the capsule biosynthetic locus of *Pasteurella multocida* M1404 (B:2). Vet Microbiol 72:121–134

Boyce JD, Harper M, St Michael F, John M, Aubry A, Parnas H, Logan SM, Wilkie IW, Ford M, Cox AD, Adler B (2009) Identification of novel glycosyltransferases required for assembly of the *Pasteurella multocida* A:1 lipopolysaccharide and their involvement in virulence. Infect Immun 77:1532–1542

Boyce JD, Harper M, Wilkie I, Adler B (2010) *Pasteurella*. In: Gyles CL, Prescott JF, Songer G, Thoen CO (eds) Pathogenesis of bacterial infections in animals. Blackwell Publishing, Iowa

Carter GR (1958) Some characteristics of type A strains of *Pasteurella multocida*. Brit Vet J 114:356–357

Carter GA (1967) Pasteurellosis: *Pasteurella multocida* and *Pasteurella haemolytica*. Adv Vet Sci Comp Med 11:321–379

Chung JY, Zhang YM, Adler B (1998) The capsule biosynthetic locus of *Pasteurella multocida* A-1. FEMS Microbiol Lett 166:289–296

Chung JY, Wilkie I, Boyce JD, Townsend KM, Frost AJ, Ghoddusi M, Adler B (2001) Role of capsule in the pathogenesis of fowl cholera caused by *Pasteurella multocida* serogroup A. Infect Immun 69:2487–2493

DeAngelis PL, Jing W, Drake RR, Achyuthan AM (1998) Identification and molecular cloning of a unique hyaluronan synthase from *Pasteurella multocida*. J Biol Chem 273:8454–8458

DeAngelis P, Gunay N, Toida T, Mao W, Linhardt R (2002) Identification of the capsular polysaccharides of type D and F *Pasteurella multocida* as unmodified heparin and chondroitin, respectively. Carbohydr Res 337:1547–1552

DeAngelis PL, White CL (2002) Identification and molecular cloning of a heparosan synthase from *Pasteurella multocida* type D. J Biol Chem 277:7209–7213

DeAngelis PL, White CL (2004) Identification of a distinct, cryptic heparosan synthase from *Pasteurella multocida* types A, D, and F. J Bacteriol 186:8529–8532

Dziva F, Muhairwa AP, Bisgaard M, Christensen H (2008) Diagnostic and typing options for investigating diseases associated with *Pasteurella multocida*. Vet Microbiol 128:1–22

Fernandez de Henestrosa AR, Badiola I, Saco M, Perez de Rozas AM, Campoy S, Barbe J (1997) Importance of the *galE* gene on the virulence of *Pasteurella multocida*. FEMS Microbiol Lett 154:311–316

Fuller TE, Kennedy MJ, Lowery DE (2000) Identification of *Pasteurella multocida* virulence genes in a septicemic mouse model using signature-tagged mutagenesis. Microb Pathog 29: 25–38

Glorioso JC, Jones GW, Rush HG, Pentler LJ, Darif CA, Coward JE (1982) Adhesion of type A *Pasteurella multocida* to rabbit pharyngeal cells and its possible role in rabbit respiratory tract infections. Infect Immun 35:1103–1109

Harper M, Boyce JD, Wilkie IW, Adler B (2003) Signature-tagged mutagenesis of *Pasteurella multocida* identifies mutants displaying differential virulence characteristics in mice and chickens. Infect Immun 71:5440–5446

Harper M, Cox AD, St Michael F, Wilkie I, Adler B, Boyce JD (2004) A heptosyltransferase mutant of *Pasteurella multocida* produces a truncated lipopolysaccharide structure and is attenuated in virulence. Infect Immun 72:3436–3443

Harper M, Boyce JD, Cox AD, St Michael F, Wilkie IW, Blackall PJ, Adler B (2007a) *Pasteurella multocida* expresses two lipopolysaccharide glycoforms simultaneously, but only a single form is required for virulence: identification of two acceptor-specific heptosyl I transferases. Infect Immun 75:3885–3893

Harper M, Cox A, St Michael F, Parnas H, Wilkie I, Blackall PJ, Adler B, Boyce JD (2007b) Decoration of *Pasteurella multocida* lipopolysaccharide with phosphocholine is important for virulence. J Bacteriol 189:7384–7391

Harper M, Cox AD, St Michael F, Ford M, Wilkie IW, Adler B, Boyce JD (2010) Natural selection in the chicken host identifies Kdo kinase residues essential for phosphorylation of *Pasteurella multocida* LPS. Infect Immun 78:3669–3677

Harper M, Cox AD, Adler B, Boyce JD (2011a) *Pasteurella multocida* lipopolysaccharide: the long and the short of it. Vet Microbiol 153:109–115

Harper M, St Michael F, John M, Vinogradov E, Adler B, Boyce JD, Cox AD (2011b) *Pasteurella multocida* Heddleston serovars 1 and 14 express different lipopolysaccharide structures but share the same lipopolysaccharide biosynthesis outer core locus. Vet Microbiol 150:289–296

Harper M, St Michael F, Vinogradov E, John M, Boyce JD, Adler B, Cox AD (2011c) Characterization of the lipopolysaccharide from *Pasteurella multocida* Heddleston serovar 9; identification of a proposed bi-functional dTDP-3-acetamido-3,6-dideoxy-α-D-glucose biosynthesis enzyme. Glycobiology 22:332–344

Heddleston KL, Rebers PA (1972) Fowl cholera. Cross immunity induced in turkeys with formalin-killed in vivo-propagated *Pasteurella multocida*. Avian Dis 16:578–586

Jacques M, Bélanger M, Diarra MS, Dargis M, Malouin F (1994) Modulation of *Pasteurella multocida* capsular polysaccharide during growth under iron restricted conditions and in vivo. Microbiology 140:263–270

Jaglic Z, Kucerova Z, Nedbalcova K, Hlozek P, Bartos M (2004) Identification of *Pasteurella multocida* Serogroup F isolates in rabbit. J Vet Med 51:467–469

Kane TA, White CL, DeAngelis PL (2006) Functional characterization of PmHS1, a *Pasteurella multocida* heparosan synthase. J Biol Chem 281:33192–33197

Muniandy N, Edgar J, Woolcock JB, Mukkur TKS (1992) Virulence, purification, structure and protective potential of the putative capsular polysaccharide of *Pasteurella multocida* 6:B. In: International workshop on pasteurellosis in production animals, Bail, Indonesia

Pandit KK, Smith JE (1993) Capsular hyaluronic acid in *Pasteurella multocida* type A and its counterpart in type D. Res Vet Sci 54:20–24

Pruimboom IM, Rimler RB, Ackermann MR, Brogden KA (1996) Capsular hyaluronic acid-mediated adhesion of *Pasteurella multocida* to turkey air sac macrophages. Avian Dis 40: 887–895

Pruimboom IM, Rimler RB, Ackermann MR (1999) Enhanced adhesion of *Pasteurella multocida* to cultured turkey peripheral blood monocytes. Infect Immun 67:1292–1296

Raetz CR, Whitfield C (2002) Lipopolysaccharide endotoxins. Annu Rev Biochem 71:635–700

Rimler RB (1994) Presumptive identification of *Pasteurella multocida* serogroups A, D and F by capsule depolymerisation with mucopolysaccharides. Vet Rec 134:191–192

Rimler RB, Rhoades KR (1987) Serogroup F, a new capsule serogroup of *Pasteurella multocida*. J Clin Microbiol 25:615–618

St Geme JW (2002) Molecular and cellular determinants of non-typeable *Haemophilus influenzae* adherence and invasion. Cell Microbiol 4:191–200

St Michael F, Li J, Cox A (2005a) Structural analysis of the core oligosaccharide from *Pasteurella multocida* strain X73. Carbohydr Res 340:1253–1257

St Michael F, Li J, Vinogradov E, Larocque S, Harper M, Cox AD (2005b) Structural analysis of the lipopolysaccharide of *Pasteurella multocida* strain VP161: identification of both Kdo-P and Kdo–Kdo species in the lipopolysaccharide. Carbohydr Res 340:59–68

St Michael F, Vinogradov E, Li J, Cox AD (2005c) Structural analysis of the lipopolysaccharide from *Pasteurella multocida* genome strain Pm70 and identification of the putative lipopolysaccharide glycosyltransferases. Glycobiology 15:323–333

St Michael F, Harper M, Parnas H, John M, Stupak J, Vinogradov E, Adler B, Boyce JD, Cox AD (2009) Structural and genetic basis for the serological differentiation of *Pasteurella multocida* Heddleston serotypes 2 and 5. J Bacteriol 191:6950–6959

St Michael F, Vinogradov E, Cox AD (2011) Structural analyses of the core oligosaccharide from the lipopolysaccharide of bovine and ovine strains of *Mannheimia haemolytica* serotype 2. Carbohydr Res 346:1333–1336

Steen JA, Steen JA, Harrison P, Seemann T, Wilkie I, Harper M, Adler B, Boyce JD (2010) Fis is essential for capsule production in *Pasteurella multocida* and regulates expression of other important virulence factors. PLoS Pathog 6:e1000750

Steenbergen SM, Lichtensteiger CA, Caughlan R, Garfinkle J, Fuller TE, Vimr ER (2005) Sialic acid metabolism and systemic pasteurellosis. Infect Immun 73:1284–1294

Tatum FM, Tabatabai LB, Briggs RE (2009) Sialic acid uptake is necessary for virulence of *Pasteurella multocida* in turkeys. Microb Pathog 46:337–344

Townsend KM, Boyce JD, Chung JY, Frost AJ, Adler B (2001) Genetic organization of *Pasteurella multocida cap* loci and development of a multiplex capsular PCR typing system. J Clin Microbiol 39:924–929

Turner J, Cho Y, Dinh NN, Waring AJ, Lehrer RI (1998) Activities of LL-37, a cathelin-associated antimicrobial peptide of human neutrophils. Antimicrob Agents Chemother 42: 2206–2214

Varga Z, Sellyei B, Magyar T (2007) Phenotypic and genotypic characterisation of *Pasteurella multocida* strains isolated from pigs in Hungary. Acta Vet Hung 55:425–434

Wu J-R, Shien J-H, Shieh HK, Chen C-F, Chang P-C (2007) Protective immunity conferred by recombinant *Pasteurella multocida* lipoprotein E (PlpE). Vaccine 25:4140–4148

Pasteurella Multocida and Immune Cells

Katharina F. Kubatzky

Abstract *Pasteurella multocida* was first discovered by Perroncito in 1878 and named after Louis Pasteur who first isolated and described this Gram-negative bacterium as the cause of fowl disease in 1880. Subsequently, *P. multocida* was also found to cause atrophic rhinitis in pigs, haemorrhagic septicaemia in cattle and respiratory diseases in many other animals. Among other factors such as lipopolysaccharide, outer membrane proteins and its capsule, the protein toxin (PMT) of *P. multocida* is an important virulence factor that determines the immunological response of the host's immune system. However, the exact molecular mechanisms taking place in cells of the innate and adaptive immune system are largely unknown for any of these virulence factors. Due to the obvious function of PMT on cells of the porcine skeletal system where it causes bone destruction, PMT was regarded as an osteolytic protein toxin. However, it remained unclear what the actual benefit for the bacteria would be. Recently, more attention was drawn to the osteoimmunological effects of PMT and the interplay between bone and immune cells. This review summarises the knowledge of effects of *P. multocida* virulence factors on the host's immune system.

Abbreviations

PMT	*Pasteurella multocida* toxin
IL	Interleukin
NFκB	Nuclear factor κB
CD	Cluster of differentiation

K. F. Kubatzky (✉)
Department of Infectious Diseases, Medical Microbiology and Hygiene,
University Hospital Heidelberg, Im Neuenheimer Feld 324,
69120 Heidelberg, Germany
e-mail: kubatzky@uni-heidelberg.de

Current Topics in Microbiology and Immunology (2012) 361: 53–72
DOI: 10.1007/82_2012_204

Published Online: 20 March 2012

Contents

1 Introduction

1.1 How the Immune System Fights Bacterial Infections

The immune system is a highly regulated network of cellular interactions that protects the organism from potential dangers that can be caused by the entrance of non-self particles such as bacteria, viruses and fungi. Haematopoietic stem cells in the bone marrow give rise to two lineages; the lymphoid lineage, which consists of B cells, T cells and NK cells and the myeloid lineage. B and T lymphocytes take part in the adaptive immune system. B cells play a role in humoral immunity as they give rise to plasma cells which produce antibodies against soluble antigens. Binding of an antigen to the B cell receptor initiates maturation to an antibody-producing plasma cell. T cells on the other hand differentiate in the bone marrow as well, but later on reside in the thymus where they need to undergo a selection process in order to fulfil their function in cell-based immunity, including recognition of antigens presented by antigen-presenting cells (APC).

The myeloid lineage can be further subdivided into two main branches, one originating from a granulocyte/monocyte precursor and the other from erythroid/megakaryocyte precursor cells. The latter differentiate into erythrocytes and platelets of the blood system. Several important immune cells are differentiated from the granulocyte/monocyte precursor cell: granulocytes, consisting of basophils, eosinophils and neutrophils, monocytes that further differentiate into macrophages and dendritic cells, and mast cells. Each of these immune cells has a specific task in the detection and clearing of an invading pathogen. Apart from their ability to act as effector cells, these cells also have the ability to communicate with each other in order to coordinate an efficient immune response.

In the presence of an infection, granulocytes are generally the first cells recruited to the site of the infection by stimuli from chemoattractants such as chemokines from the inflamed tissue or bacterial formyl peptides released by the invading microorganism itself (Muller 2011). They will then, at the site of infection, transmigrate through the epithelium and invade the inflamed tissue.

Granulocytes are effector cells that produce bactericidal substances and can phagocytose bacteria. Neutrophils also assist in maintaining this local inflammatory response through the production of reactive oxygen species (ROS), proteases and antimicrobial peptides (Stuart and Ezekowitz 2005). This then attracts macrophages, localised close to the inflamed tissue to the site of infection.

Digestion of pathogens through phagocytosis allows the subsequent presentation of the digested peptides via major histocompatibility complex II (MHC II) to T cells, a process that bridges the innate and the adaptive immune system. Phagocytosis involves the uptake of microbial particles through formation of a phagosome and the subsequent fusion with a lysosome to the phagolysosome in order to degrade the ingested material. The lysosome is a low pH milieu and contains bactericidal substances. Additionally, ROS as well as NO are produced that kill the pathogen. This process is often initiated by the complement cascade system that resides in the plasma. In the classical pathway, specific antibodies recognise an antigen, bind to it and thus coat the bacterium, which renders it more visible for phagocytes. In the alternative pathway, complement protein directly binds to the microbial pathogen. This opsonisation of bacteria is recognised by complement receptors on immune cells, which activates them to phagocytose the opsonised invader.

Phagocytosis is also associated with dendritic cells that can leave the tissue and migrate to regional lymph nodes. Once they are activated, dendritic cells are the most powerful antigen presenting cell type. Depending on the antigen detected, immature dendritic cells differentiate under the influence of paracrine cytokine stimulation to mature dendritic cells and are capable of inducing the polarisation of naïve T-cells (Sabatte et al. 2007). Polarisation of naïve T-cells into the helper T cells (Th) Thl or Th2 depends on the kind of costimulatory molecules and cytokines that are produced by myeloid dendritic cells (Guermonprez et al. 2002). Cytokines are important mediators of the cellular immune response and can act in an autocrine as well as in a paracrine fashion. Cytokines play a predominant role in haematopoiesis and also affect the regulation of the immune response of mature haematopoietic cells. Many cytokines have a proliferative function that allows the immune system to specifically produce more effector cells in response to the activation of a target cell. In addition, the cytokine pattern released by innate immune cells stimulates naïve T cells to develop into helper T cells and determines the ultimate immune response given by these cells. While Th1 cells secrete IL-2, IFN-γ, TNF-α and participate in cell-mediated immune responses, Th2 cells are important helper cells for B cells through the production of IL-4, IL-5, IL-6, IL-10.

Immune cells are known to express a variety of so-called pattern recognition receptors that specifically recognise pathogen-associated molecular patterns (PAMPs). Recognition and binding of a microbial stimulus leads to the subsequent activation of intracellular signalling cascades, eventually resulting in clearance of the threat. Innate immune receptors can be transmembrane receptors such as the toll-like receptors (TLRs), as well as soluble components that mark microbial substances for detection by complement receptors, integrins or Fc receptors (Underhill 2003). TLRs are pattern recognition receptors as they are not activated by specific ligands, but through recognition of more general pathogen-associated PAMPs such as lipids (lipopolysaccharides and lipoproteins), proteins (flagellin) or DNA and double-stranded RNA (Akira et al. 2006). The expression pattern of these receptors depends on the type of immune cell and their expression is not restricted to the haematopoietic system (Anders et al. 2004). While TLRs 1, 2, 4, 5 and 6 are primarily located at the cell surface to detect microbial PAMPs, TLRs 3, 7, 8 and 9 are localised in endosomal compartments which allows efficient recognition of viral or bacterial nucleic acids. Eventually, TLR signalling events result in the activation of transcription factors such as NFκB, Map kinases or IRFs (Brown et al. 2011). A central task of TLR signalling is the NFκB-dependent production of inflammatory cytokines as an initial defence mechanism. However, novel functions are emerging and TLRs were shown to play a substantial role in phagocytosis, antigen presentation and crosstalk with T cells as well as clearance of microbes through autophagy (Blander and Medzhitov 2004).

1.2 Bacterial Strategies of Immune Evasion

Many bacteria live in close relationships with their host and sometimes they even benefit the host. They have thus developed strategies in a co-evolutionary process to overcome the above-mentioned arsenal of mechanisms that help the immune system fight the microbe. Interestingly, many commensals can turn into pathogens, for example upon a change in the health of the host or due to changes in the pathogenicity of the microbe. The word pathogenicity describes the sum of all the characteristics of a microorganism that play a role in infection. The term virulence on the other hand describes the degree of pathogenicity of a specific microorganism. There are for example several strains of a given microorganism that due to the absence of virulence factors have a different degree of virulence. These virulence factors can be produced as an integral part of the bacterium, such as its cell wall or it can be synthesised upon a certain trigger and then cause an infection. In general it can be said that the pathogenicity of a microorganism depends on its capability to adhere to host cells, to colonise at the site of infection and then to further invade the organism. Additionally, bacteria can produce toxins that help to destroy host cells.

Strategies to evade the host's immune response can therefore include the manipulation of the immune system, for example through targeting specific signalling pathways that result in dampening the immune response, or by changes in the bacterium itself or its localisation (Srikumaran et al. 2008).

The extracellular matrix is an important interaction platform between bacteria and their host. It therefore participates in many mechanisms of virulence and evasion (Fig. 1). Bacterial surface components help the microorganism in the process of bacterial adhesion and invasion but on the other hand provide characteristic molecular patterns that can be recognised by immune cells. Factors that help the bacteria to adhere and to prevent being cleared from the mucosa or

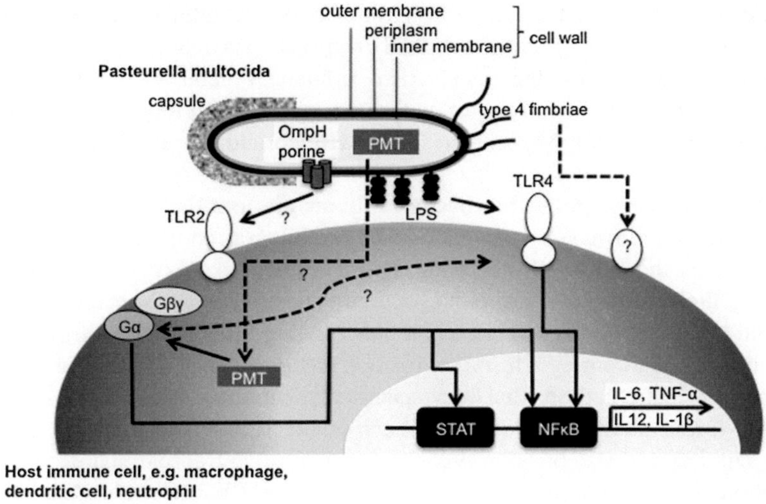

Fig. 1 *Strategies of bacterial host colonisation.* Gram-negative bacteria such as *P. multocida* express various virulence factors that determine their pathogenicity. The most important virulence factors of *P. multocida* are the presence of a polysaccharide capsule, fimbriae, lipopolysaccharide and outer membrane proteins, e.g. the porin OmpH. PMT is an AB toxin and probably encoded by a bacteriophage. PMT acts intracellularly and deamidates G proteins eventually causing the activation of signal transducer and activator of transcription factors (STAT) or NFκB. Toll-like receptors present on the host's immune cells detect LPS (TLR4) and probably also proteins of the outer membrane. TLR signalling cascades result in the activation of the transcription factor NFκB and the subsequent release of inflammatory cytokines. Known interactions are depicted using *solid arrows*, hypothetical pathways are shown using *dotted arrows*

epithelium include lectins, fimbriae, as well as non-fimbrial adhesins and glycosaminoglycan-binding proteins.

To avoid recognition through the host, bacteria have developed ways to alter their surface structures through antigenic hypervariability of structures that are easily detected by the immune system (Finlay and McFadden 2006). Factors that can undergo variation include the carbohydrates that are used to build up a capsule surrounding and masking the bacterial cell wall, pili and proteins being expressed in the outer membrane (Omp). Lipopolysaccharide (LPS), a typical constituent of the outer membrane of the cell wall of Gram-negative bacteria that causes the typical symptoms of a bacterial infection such as fever, is varied for example through addition of side groups causing the development of different serotypes of a bacterial strain. This variation can affect antigenicity, serum sensitivity but also adhesion (van der Woude and Baumler 2004).

Additionally, bacteria protect themselves by attacking parts of the complement system or immunoglobulins through specifically acting bacterial proteases. Even if these mechanisms fail and the pathogen is phagocytosed, bacteria have means to act intracellularly, for example by inhibiting phagolysosomal fusion (Finlay and

McFadden 2006; Frehel et al. 1986). Subversion of intracellular pathways by toxins and peptides or proteins released from bacteria has come to attention in recent years. Eventually this mechanism influences gene transcription through post-translational modifications of cellular targets or inhibition of degradation processes (Finlay and McFadden 2006; Hildebrand et al. 2010; Ribet and Cossart 2010).

Strains capable of producing bacterial toxins are usually more virulent. While the word endotoxin means substances that are integral components of the bacterium, such as the cell wall, bacterial DNA or RNA, exotoxins are proteins produced by the bacteria that are either actively secreted or will be released during bacterial lysis. AB toxins contain a receptor-binding domain that facilitates uptake of the bacterial protein by a host cell surface receptor. Bacterial protein toxins can have an enzymatic activity that very specifically modifies intracellular host target proteins. Often their targets are members of the Rho GTPase family (Aktories 2011), but they can also be other proteins that play a central role in host viability or regulation of the immune system (Lax and Thomas 2002; Oswald et al. 2005).

2 *Pasteurella multocida* and Immune Cells

2.1 Pasteurella multocida-Associated Diseases

Pasteurella multocida is a Gram-negative, facultatively anaerobic coccobacillus that can be found on many animals worldwide where it can cause a variety of infections. The *P. multocida* species can be further subdivided into four subspecies, namely multocida, gallicida, septica and tigris (Harper et al. 2006). Serologically, five types (A, B, D, E and F) can be defined according to differences of the capsular antigen (Arumugam et al. 2011; Carter 1955) and 16 types based on differences in LPS composition (Harper et al. 2011; Heddleston et al. 1972). Although *P. multocida* infections are zoonotic diseases, humans are rarely affected and bacteria are usually transferred from animals for example through cat or dog bites and scratches. For humans it can be considered an opportunistic infection as it is mainly found in immune-compromised patients. This can also be true for animals, where *P. multocida* is usually a commensal that can become pathogenic due to an underlying infection or changes in the expression pattern of virulence factors. Clinically, *P. multocida* infections in humans are mostly associated with skin and soft tissue infections. However, respiratory tract infections as well as bone and joint infections are also known and the latter can result in osteomyelitis and septic arthritis (Zurlo 2005).

Pasteurella multocida infections such as fowl cholera, bovine haemorrhagic septicaemia, enzootic calf pneumonia, swine atrophic rhinitis or rabbit snuffles are economically relevant problems and the mechanisms of pathogenesis have therefore been extensively studied in various animal models. Details of these disease syndromes are described in "Host-Pathogen Interactions.". Interestingly, different serotypes are associated with specific types of diseases. While fowl cholera is associated with serotype A:1 and A:3 strains (Adler et al. 1999), haemorrhagic septicaemia is associated with B:2 and E:2 strains (Shivachandra et al. 2011) while atrophic rhinitis of pigs is primarily caused by toxigenic type D and A strains (Amigot et al. 1998).

A number of virulence factors have so far been defined and will be discussed below, but the molecular basis for the interference between *P. multocida* strains and the host's immune system is only incompletely understood. Thus the application of new methods such as proteomic analysis, more refined detection methods and the availability of the complete genomic sequence of *P. multocida* (May et al. 2001; "The Key Surface Components of *Pasteurella multocida*: Capsule and Lipopolysaccharide") are likely to enable us to define other factors that contribute to the disease or are important factors in the interaction between host and pathogen.

2.2 Major Virulence Factors of Pasteurella multocida

2.2.1 LPS

LPS is a component of the bacterial cell wall typical for Gram-negative pathogens. Detection of LPS by TLR4 expressed on cells of the innate immune system ultimately leads to a humoral response characterised by the secretion of antibodies. Due to its prominent role, LPS acts as both a virulence factor and a serovar-specific immunogen. Of the 16 different serotypes known for *P. multocida* (Heddleston et al. 1972), at least some are genetically related, meaning that they share the same outer core locus. Distinct point mutations in one of the enzymes involved in attachment of sugar modifications such as the phosphoethanolamine (PEtn) transferase gene or phosphocholine synthase generate a modified LPS (Harper et al. 2011). Changes in the outer core of the LPS oligosaccharides can be the result of bacterial phase-variation strategies; however, this has not been shown for *P. multocida* (Harper et al. 2007). An unusual finding was, however, that *P. multocida* strains simultaneously express two LPS glycoforms, which differ in their inner core (Harper et al. 2007). This highlights the fact that detailed knowledge of the genetic background and structural studies will be a prerequisite in order to identify suitable targets to generate efficient vaccines. Details of LPS structure are described in "Pathogenomics of *Pasteurella multocida*".

Recognition of LPS occurs via binding to TLR4 and eventually results in the release of NFκB-induced inflammatory cytokines and expression of adhesion molecules that attract leukocytes to the site of infection. The work of Galdiero et al. (2000) showed that whole *P. multocida* bacteria as well as LPS preparations isolated from *P. multocida* enabled efficient adhesion of neutrophils to isolated bovine endothelial cells, presumably supported by the upregulation of adhesion markers such as VCAM-1 (Richardson et al. 1997) and followed by subsequent transmigration through the endothelial cell layer 2–4 h after stimulation. *P. multocida* LPS was also able to induce the expression and the subsequent release of proinflammatory and immunomodulatory cytokines such as IL-1α, IL-6, TNF-α, IFN-γ and IL-12 from a mix of murine splenocytes (Iovane et al. 1998). This result is in line with the many reports using *E. coli* or *Salmonella* LPS as a stimulating agent to induce TLR4-dependent signalling cascades; however, no data are available that would directly compare *P. multocida* LPS with commercially available LPS preparations or evaluate different *P. multocida* LPS compositions for TLR4 signalling.

The endotoxin LPS is an exogenous pyrogen and LPS-initiated release of cytokines eventually induces fever. Whether the endotoxic activity of *P. multocida* LPS is comparable to that of other microbes is controversially discussed in the available literature and the susceptibility might be species specific (Dabo et al. 2007; Harper et al. 2006). However, from studies employing bacteria with mutations in the sugar core of LPS or mutants with truncated LPS, it is also clear that LPS is required to cause disease, as these mutant strains show decreased viability in animals (Fernandez de Henestrosa et al. 1997; Harper et al. 2004).

From the side of the host, LPS can generally be considered a protective antigen that can be detected efficiently. It has therefore been tested whether *P. multocida* LPS might also be a good immunogen. Passive immunisation using monoclonal antibodies against LPS from a serotype A strain indeed killed bacteria and protected mice from infection. However, this was true only for homologous infections with the same mice strain, suggesting that protection is serovar specific (Wijewardana and Sutherland 1990). Other examples include an opsonic monoclonal antibody that was able to deliver opsonized bacteria for phagocytosis by macrophages. This was demonstrated in an experimental setting were peritoneal macrophages were prepared and incubated with *P. multocida*, monoclonal antibodies against LPS and complement to facilitate uptake of bacteria (Ramdani and Adler 1991). However, passive immunisation of mice using affinity-purified anti-LPS serum did not protect mice efficiently against infection, thus suggesting that LPS plays a minor role as a protecting immunogen (Lu et al. 1991). One method to enhance the effect of LPS antibodies is the use of an LPS-protein mix or ribosomes, which achieved complete protection in a mouse model system under homologous infection conditions (Dabo et al. 2007; Rimler and Phillips 1986; Ryu and Kim 2000).

2.2.2 Capsule

The antigenicity of the capsule is used to identify the five known serogroups A, B, D, E and F (Carter 1955), suggesting that the capsule gives rise to B cell-mediated antibody production, although this notion remains controversial. Details of capsule composition, biosynthesis and regulation and role in pathogenesis are presented in "Pathogenomics of *Pasteurella multocida*".

2.2.3 Adhesins

Pili or type IV fimbriae are widespread among Gram-negative bacteria but can also be found in Gram-positive bacteria (Proft and Baker 2009). They are built up by long filaments consisting of app. 20 kDa fimbrial subunits that facilitate adhesion to host cells. Initially, adherence occurs at epithelial cells and this represents a crucial step in successful colonisation of the host and subsequent infection. While epithelial cells do not belong to haematopoietic immune cells, they present an important physical barrier with effective antimicrobial defence mechanisms through production of antimicrobial peptides and the activation of innate immune receptors (Marques and Boneca 2011). Although type I fimbriae were recently shown to be recognised by TLR4 proteins expressed on macrophages (Mossman et al. 2008), nothing is known about type IV fimbriae and the resulting response of the innate immune system.

Pili are important virulence factors that facilitate colonisation of the host; on the other hand, type IV fimbriae have been used successfully as immunogenic targets to develop vaccines, e.g. against *Yersiniae pestis*, *Salmonella enteriditis*, *Pseudomonas aeruginosa*, *Dichelobacter nodosus* and *Moraxella bovis* (Adler et al. 1999; Proft and Baker 2009). The fimbrial subunit protein of *P. multocida*, PtfA, is a 15 kDa protein with a high degree of similarity to other fimbrial proteins, except for its prolonged signal sequence (Adler et al. 1999; Doughty et al. 2000). The protein is found on serotype A, B and D strains (Ruffolo et al. 1997) and is often associated with increased virulence (Harper et al. 2006). Due to the significant divergence in the C-terminal part between various *P. multocida* strains, development of a vaccine might be difficult (Doughty et al. 2000). There are also other *P. multocida* proteins with similarity to proteins encoding for fimbriae, fibrils or filamentous haemagglutinins in other bacteria, but no details are available concerning their interaction with immune cells.

2.2.4 Outer Membrane Proteins and Porins

To date, some outer membrane proteins (Omps) of *P. multocida*, including OmpH, OmpA, P6-like protein, PlpB, GlpQ, Lpp and Oma78 have been characterised experimentally concerning their structure and immunogenicity; recently, proteomic studies revealed the existence of 28 other proteins and ten of these are predicted to be porins (Boyce et al. 2006). Several Omps were shown to act as virulence factors with OmpH being the best characterised among them. It was found to be present in basically all bovine isolates (Dabo et al. 2007) and is discussed as a vaccine candidate as it was shown to protect mice and chicken against *P. multocida* challenge (Luo et al. 1999; Vasfi Marandi and Mittal 1997).

OmpH is a trimeric porin that is also known as Protein H, with a molecular mass of 37.5 kDa (Chevalier et al. 1992). So far it is the only porin that has been tested in more detail for its ability to stimulate immune cells. In a study comparing the effects of *P. multocida* LPS with recombinant and reconstituted porin, OmpH was shown to modulate the response of the immune system by inducing the expression and production of proinflammatory cytokines such as IL-1α, IL-6, TNF-α or IFN-γ and IL-12 from monocytes and macrophages of murine splenocytes (Iovane et al. 1998). None of the typical B cell-produced Th2 cytokines such as IL-4 and IL-10 were detected. It can therefore be hypothesised that OmpH does not directly influence the adaptive immune system but that it can influence the secretion of cytokines by innate immune cells. The release of the cytokines was delayed compared to LPS and the observed gene induction was less efficient, at least at low concentrations. It is currently unknown how *P. multocida* porins and OmpH interact with the host cell. Porins from other Gram-negative microorganisms however, were shown to be detected by TLR2 receptors (Wetzler 2010). Why this would cause differences in the kinetics of cytokine production compared to the LPS-mediated initiation of TLR4 signalling is unclear.

Bacteria can act as chemoattractants and it was therefore tested whether OmpH alone was sufficient to trigger neutrophil migration (Galdiero et al. 1998). Indeed, the presence of porin-induced neutrophil migration and actin polymerisation in a concentration-dependent manner, eventually resulting in an oxidative burst (Galdiero et al. 1998). As *P. multocida* suspensions induced the adhesion of neutrophils to endothelial cells, it was tested whether the same could be achieved using isolated membrane components (Galdiero et al. 2000). Adhesion could be induced using either 200 nM of isolated LPS or 50–200 nM Protein H and both substances were able to stimulate bacterial transmigration; however, maximum transmigration for porin was reached at a later time point. Experiments investigating the ability of OmpH to trigger the production of the immunoregulatory messenger molecule NO from murine macrophages showed that the porin itself was unable to do so. However, the presence of IFN-γ increased the induction of iNOS (inducible nitric oxide synthase) expression and subsequent NO release from macrophages compared to the stimulation with IFN-γ alone (Marcatili et al. 2000).

2.2.5 *Pasteurella multocida* Toxin

Toxigenic *P. multocida* serotype A and D strains are able to produce a 146 kDa protein toxin (PMT). In swine, these toxigenic strains are widespread and the reduced weight gain of animals presents an economically relevant problem (Foged 1992; Harper et al. 2006). PMT is the causative agent of atrophic rhinitis, which causes osteoclastic bone resorption at the nasal turbinates (Felix et al. 1992) and isolated PMT is able to induce all the symptoms of the disease (Chrisp and Foged 1991; Dominick and Rimler 1986). On the genetic level, the toxA gene sequence encoding for PMT has a lower G + C content compared to the rest of the *Pasteurella* genome, suggesting that it is phage-encoded (Pullinger et al. 2004). In addition, no specific sequence provides for a mechanism to secrete PMT. Using electron microscopy, PMT was found to reside in the cytoplasm and to be absent in undamaged bacteria (iDali et al. 1991). Therefore, it is possible that PMT is only released from the cells when the temperate phage is induced (Pullinger et al. 2004).

In vitro, PMT acts as a potent mitogen for a variety of cell types such as fibroblasts, bladder epithelial cells and osteoclasts (Hoskins et al. 1997; Martineau-Doize et al. 1993; Rozengurt et al. 1990). PMT is taken up by host cells through receptor-mediated endocytosis, presumably via binding to gangliosides (Dudet et al. 1996; Pettit et al. 1993). The heterotrimeric G proteins $G\alpha_q$, $G\alpha_{13}$, and $G\alpha_i$ (Orth et al. 2005, 2008; Wilson et al. 1997; Zywietz et al. 2001), which PMT renders constitutively active through deamidation of a conserved glutamine residue to glutamate (Orth et al. 2009), are the specific cellular targets of PMT. This leads to the activation of intracellular signalling pathways such as Rho GTPase signalling, phospholipase Cβ (PLCβ), activation of mitogen activated protein (Map) kinases, the Janus kinase-signal transducer and activator of transcription (JAK-STAT) pathway or the activation of phosphoinositol (PI) 3 kinases (Wilson and Ho 2011). PMT-modulated signalling cascades have been identified using cell lines and not primary immune cells, thus it is still under investigation whether these signalling changes eventually modify the immune response of the host.

One major pathway of haematopoietic cells including immune cells is the activation of JAK-STAT signalling. As PMT is able to activate this pathway (Orth et al. 2007), it can be hypothesised that the toxin might have a function in immune signalling. Dendritic cells are the most potent antigen-presenting cells and the only APC capable of initiating immune responses of the adaptive immune system. Therefore, several groups have investigated the effect of PMT on these cells (Bagley et al. 2005; Blocker et al. 2006). Maturation of human monocytes into dendritic cells is monitored measuring the expression of specific surface markers, such as CD80, 83, 86 and HLA-DR. The PMT-mediated upregulation of these markers was concentration dependent and detectable at 10 pM, but was highest at 7–20 nM. Bacterial protein toxins such as cholera toxin, pertussis toxin, heat-labile enterotoxin (*E. coli*) or adenylate cyclase (*Bordetella pertussis*) have been shown to trigger monocyte maturation through elevation of the intracellular cAMP levels either directly or indirectly through ADP-ribosylation of G proteins and subsequent constitutive adenylate cyclase activation (Bagley et al. 2002a, b). Additionally, the production of IL-12, which induces proliferation and cytokine production in T cells, is suppressed (Bagley et al. 2002a, b; Kubin et al. 1994). PMT on the other hand activates PLCβ followed by an increase in intracellular calcium levels. Interestingly, calcium is known to induce maturation of dendritic cells and to inhibit IL-12 production (Faries et al. 2001). It was therefore investigated whether PMT would have similar effects as other protein toxins on DC reactivity, though using a different pathway.

In contrast to the endotoxin LPS that is an efficient activator of cytokine secretion, due to the elevation in cAMP levels, bacterial protein toxins do not induce the production of cytokines (Zidek 1999) and are able to downregulate LPS-induced production of IL-12 (Bagley et al. 2002b; Gagliardi et al. 2000). This was also verified for PMT (Bagley et al. 2005). Nevertheless, due to the activated phenotype induced by PMT a co-culture of PMT-treated dendritic cells with naive T cells increased T cell proliferation comparable to LPS or cholera toxin treatment. The ability of toxins to induce maturation of DCs in vitro usually correlates well with their ability to boost an antibody response in vivo due to efficient antigen presentation and T cell-supported B cell activation (Pulendran and Ahmed 2011). However, when used as an antigen for vaccination, PMT was repeatedly shown to downregulate the antibody response and to even suppress the production of antibodies in response to the injection of cholera toxin (Bagley et al. 2005; Nielsen et al. 1991; van Diemen et al. 1994). These findings suggest that other mechanisms in addition to the activation and maturation of dendritic cells exist, rendering the organism unresponsive to active toxin.

The migration of dendritic cells to regional lymph nodes is an important factor in generating an efficient adaptive immune response. Using murine dendritic cells, PMT was found to induce RhoA-mediated actin polymerisation (Blocker et al. 2006). Morphologically, this translated into altered cell morphology with long needle-like extensions instead of the usual irregularly formed shape of immature dendritic cells. As a consequence, complement-induced actin re-organisation was impaired, leading to the inhibition of cell migration in response to chemoattractants such as complement C5a or the chemokines CCL5 and CCL9. In contrast to *P. multocida* factors such as LPS or Protein H localised at the bacterial cell wall, PMT itself did not act as a chemoattractant. However, measuring the uptake of the dye texas red macropinocytosis, i.e. the formation of large endocytic vesicles for the uptake of pathogens or antigens, was found unchanged. While macropinocytosis also requires an active cytoskeleton, mechanistically, Rac activation might be more important here and was shown to be unaffected by PMT (Blocker et al. 2006; Nobes and Marsh 2000).

2.2.6 Aspects of PMT in Osteoimmunology

In mammals bone cells regulate the skeleton while the immune system deals with the detection and destruction of invading pathogens. Interestingly, there is strong crosstalk between these two systems that led scientists to define the emerging field of osteoimmunology (Arron and Choi 2000; Mensah et al. 2009). Osteoclasts are multinucleated cells that break down bone. They are however generated through fusion of mononuclear precursor cells developed from macrophages via a macrophage colony-stimulating factor (M-CSF) and RANKL-mediated differentiation process and therefore belong to the haematopoietic lineage. To investigate the effects of PMT on bone marrow cells in more detail, haematopoietic murine progenitor cells were cultivated with PMT. When the ability of haematopoietic cells to take up fluorescently labelled PMT was tested, all of the cells tested took up PMT to varying degrees (Fig. 2). However, in an experimental setup to test the ability to survive without additional cytokines, only two types of cells were found to survive (Hildebrand et al. 2011) (Fig. 3a). Subsequent characterisation of the two cell types by FACS analysis showed that the surviving cells were B cells and macrophages, which, after prolonged incubation with the toxin, further differentiated into osteoclasts as shown by fluorescence staining for the osteoclast-specific enzyme tartrate-resistant acidic phosphatase (TRAP) and the presence of multiple

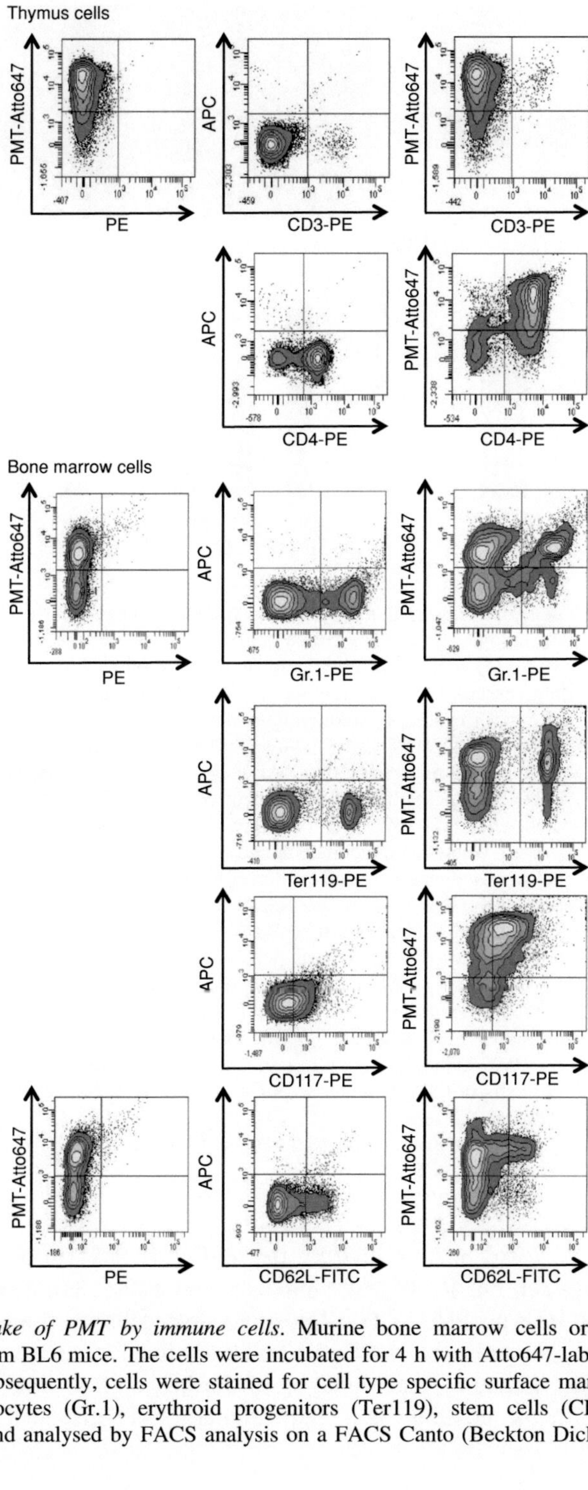

Fig. 2 *Uptake of PMT by immune cells.* Murine bone marrow cells or thymus cells were prepared from BL6 mice. The cells were incubated for 4 h with Atto647-labelled PMT (6.5 nM) at 37°C. Subsequently, cells were stained for cell type specific surface markers [T cells (CD3, CD4), monocytes (Gr.1), erythroid progenitors (Ter119), stem cells (CD117) and platelets (CD62L)] and analysed by FACS analysis on a FACS Canto (Beckton Dickinson)

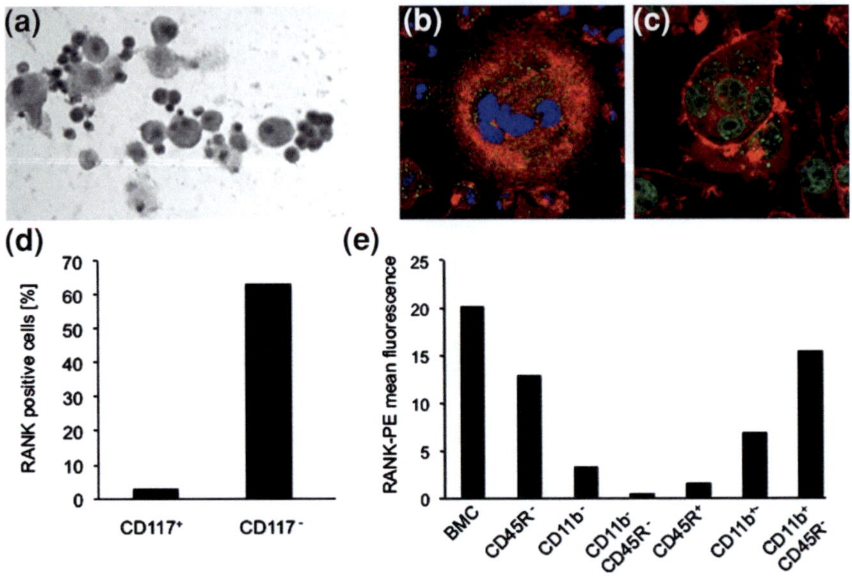

Fig. 3 *PMT induces the differentiation of murine haematopoietic bone marrow cells into B cells and macrophages.* **a** Morphological analysis of bone marrow cells (BMCs). Cytospins of BMCs stimulated with PMT (6.5 nM) were paraformaldehyde fixed, stained with hematoxylin and eosin, and mounted with glycerine-PBS for microscopy. **b** TRAP staining of BMCs differentiated with PMT (6.5 nM) for 5 days. The actin cytoskeleton was stained using Phalloidin-TRITC (*red*). Nuclei were counterstained with SYTOX Green (shown in *blue*). Osteoclasts or pre-osteoclasts were characterised by detecting the expression of TRAP using the ELF 97 phosphatase substrate (*yellow-green*). Osteoclasts additionally display multi-nuclearity. **c** TRAP staining of RAW 276.4 macrophages differentiated with PMT (6.5 nM) for 6 days. The actin cytoskeleton was stained using Phalloidin-TRITC (*red*). Nuclei were counterstained with SYTOX Green (shown in *green*). Osteoclasts or pre-osteoclasts were characterised by detecting the expression of TRAP using the ELF 97 phosphatase substrate (*yellow-green*). Osteoclasts additionally display multi-nuclearity. **d** Purification of $CD117^+$ stem cells. Stem cell receptor (CD117) expressing cells were purified from a mix of total bone marrow cells by MACS purification technology, double stained for CD117 and CD11b, and analysed by FACS. Quantification of macrophages and osteoclasts in $CD117^+$ and $CD117^-$ cells. After 9 days of stimulation with PMT (6.5 nM), the number of $RANK^+$ osteoclasts from the $CD117^+$ and $CD117^-$ fractions were quantified by FACS analysis with specific antibodies. **e** PMT-induced osteoclast generation is B-cell dependent. Using MACS cell purification of $B220^+$ cells or $CD11b^+$ cells or purification of both cell types, five different fractions (BMC mix, $CD45R^-$ $CD11b^-$, $CD45R^+$, $CD11b^+$, and $CD45R^+$ $CD11b^+$ cells) were generated. After 14 days of PMT stimulation (6.5 nM), cells were stained for the osteoclast marker RANK and analysed by FACS

nuclei (Fig. 3b). This was also verified by the ability of PMT to differentiate a murine macrophage cell line RAW 264.7 into osteoclasts (Fig. 3c). Although stem cells were able to take up PMT, CD117-positive stem cells were unable to differentiate into receptor activator of NFκB (Rank) positive osteoclasts, suggesting that PMT triggers immature macrophages to further mature into osteoclasts (Fig. 3d). While it is possible that intracellular, PMT-mediated signalling pathways are able to induce this maturation process at least for mature macrophages, subsequent experiments support the hypothesis that B cells play a central role in PMT-derived osteoclast differentiation. Generating isolated macrophages (CD11b) and B cells (CD45R) through cells MACS-assisted cell sorting, it was shown purified macrophages had a decreased ability to differentiate into osteoclasts in the presence of PMT when compared with a mix of bone marrow cells (Fig. 3e). However, the addition of purified B cells to the macrophage population restored osteoclastogenesis, suggesting that the B cell population plays an essential role in PMT-mediated differentiation of macrophages into osteoclasts. Indeed, PMT-treated B cells produce large amounts of osteoclastic cytokines such as TNF-α, IL-6, RANKL and IL-1β. However, it is also possible that direct cell–cell-contacts are also required for this process and transwell experiments would be required to investigate this question in more detail. Apart from its function as an inducer of osteoclastogenesis, IL-1β is an endogenous pyrogen that causes fever. On a molecular level, pro-IL-1β production causes a cell to activate the inflammasome, which allows the cleavage of the inactive zymogen through the cysteine protease caspase-1 into its active mature form of cleaved IL-1β (Broz and Monack 2011; Gross et al. 2011). The process of inflammasome activation is associated with the activation of apoptotic pathways (Miao et al. 2011). This however is in great contrast to our findings that PMT protects cells from apoptosis through activation of the survival kinases Pim and Akt (Preuss et al. 2010), suggesting that the cleavage of pro-IL-1β in the absence of an inflammasome activation by yet unknown mechanisms might be possible.

3 Discussion

Pasteurella multocida is a commensal microorganism for many animals and its colonisation, with the exception of toxA producing strains of serotype A and D, does usually not harm its host (Biberstein 1990). However, it can turn into a pathogen causing severe diseases. It is not well understood how the transition from commensal into a pathogen occurs, but it is possible that an underlying infection that weakens the immune system triggers the development of a secondary opportunistic infection. While in this scenario a change in the immune status of the host is the pivotal event, the ability of *P. multocida* to express a variety of virulence factors is likely to play a role as well. In particular, the continual adaption to the host through genetic exchange is likely to contribute to both successful commensalism as well as pathogenesis (Michael et al. 2012; Redfield et al. 2006). It is also conceivable that inherent differences exist between pathogenic versus commensal populations that have not yet been described (Dabo et al. 2007). With proteomic approaches becoming a technology that is used more frequently and conveniently to address issues regarding differences at the protein level, this question may be answered within the next years.

Indeed, many of the data available on *P. multocida* virulence factors were generated at a time when studies were restricted to simple experimental setups. Additionally, at that time only a limited amount of information was available regarding signalling pathways of the immune system such as the TLR pathway (Medzhitov et al. 1997). Therefore, many publications investigating the influence of *P. multocida* virulence factors were limited to a phenotypic description without providing molecular details on the mechanism of action. It will be of great interest to return to these early findings regarding the function of virulence factors such as LPS, adhesins and outer membrane proteins and to elucidate the molecular details now. Additionally, the availability of more sophisticated immunological and biochemical techniques and the feasibility of proteomic studies at the level of the host as well as of the microorganism will give us new insights into the process of mutual regulation between the two and will ultimately help to develop tools to prevent or control *P. multocida* related animal diseases. Using these techniques it will also be of interest to identify and characterise new proteins and to evaluate their function in pathogenicity. Another conundrum is the finding that although *P. multocida* is an important pathogen for many wildlife and domestic animals where it causes severe diseases, for humans *P. multocida* related infections are rare and often associated with immune compromised patients (Harper et al. 2006; Zurlo 2005). Whether this is due to differences in the immune system that allows better detection or more efficient clearing is currently unknown.

Bacteria do not only interact with cells of the immune system, but they also play an important role in the pathology of bone-related conditions. Often, a combination of inflammation and loss of bone is detected in an on-going bacterial infection. Examples of that type include periodontal disease, osteomyelitis or bacterial arthritis (Nair et al. 1996). Considering the fact that osteoclasts are differentiated from macrophages through M-CSF and RANKL-mediated differentiation process (Fig. 4a), an excess in osteoclast differentiation might dampen the innate immune response by macrophages. However, it is not clear whether bacteria cause pathological bone loss through direct destruction of bone components, for example, through manipulation of cellular processes that regulate bone resorption, or through activation of the host's immune system, eventually leading to the production of osteolytic factors and proinflammatory cytokines. A prime example of bacterially induced osteoclastogenesis is represented by PMT. While the detailed mechanism of PMT on osteoclast activity remains largely unknown, phenotypically PMT was shown to induce the differentiation of pre-osteoclasts into osteoclasts (Jutras and Martineau-Doize 1996; Martineau-Doize et al. 1993), eventually causing increased bone resorption. Additionally, PMT seems to inhibit osteoblasts which impedes effective bone regeneration (Mullan and Lax 1998).

While numerous reports discuss the possibility of a close interaction or even relationship between B-cells and osteoclasts (Blin-Wakkach et al. 2006; Calvani et al. 2004; Horowitz et al. 2005; Miyaura et al. 1997), little is known on this kind of crosstalk with respect to the influence of bacterial infections. Additionally, it is not clear why bacteria target the bone system at all and if this might represent a new mechanism of immune evasion. Due to the close connection between the bone

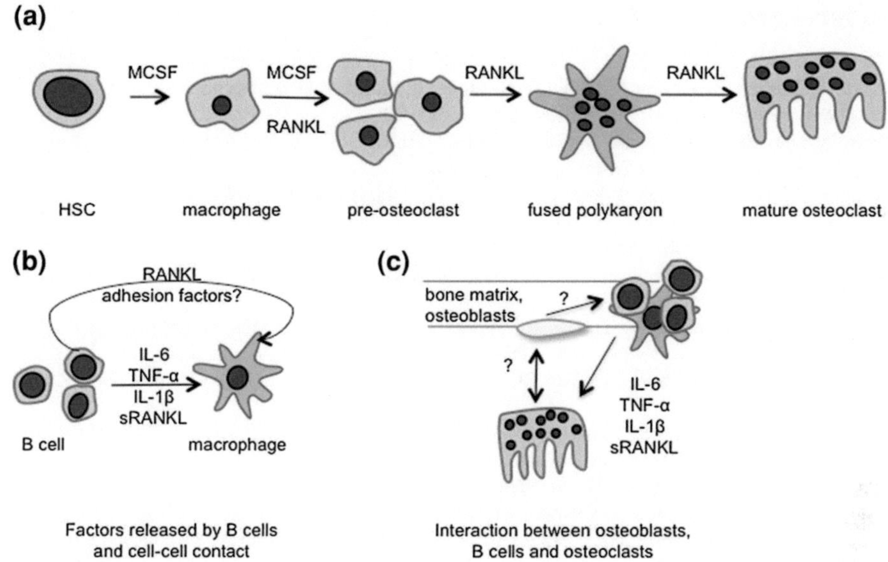

Fig. 4 *Osteoclast differentiation.* Classical osteoclastogenesis **a** involves the cytokine-induced differentiation of haematopoietic progenitors from the myeloid lineage and subsequent differentiation into macrophages and osteoclasts via the two factors M-CSF and RANKL. **b** The recently described model of PMT-induced osteoclastogenesis involves the cooperation between B cells and macrophages to efficiently generate osteoclasts (Hildebrand et al. 2011). In addition to cytokines secreted by B cells, direct cell–cell contact is likely to play a role. **c** In a hypothetical model of PMT-mediated osteoclast formation the interaction of osteoblasts and osteoblasts as well as the interaction between immune cells and cells of the bone system are likely to contribute to the differentiation process

and immune system, manipulating the response of immune cells might allow bacteria to shift the equilibrium from bone homeostasis to increased bone resorption as a side effect. Some bacterial products directly stimulate bone resorption, for example LPS, teichoic acid, lipid A associated proteins, as well as cell wall and cell surface components of a number of bacterial strains (Nair et al. 1996). Recently, we showed that PMT influences the activity of B cells and that B cells are important helper cells that stimulate PMT-mediated osteoclast differentiation from isolated murine macrophages (Hildebrand et al. 2011) (Fig. 4b). We hypothesise that PMT will be a valuable tool to investigate osteoimmunological questions and to identify signalling pathways that connect signals from osteoblasts, osteoclasts and immune cells that play a role in the regulation of osteoclast formation (Fig. 4c).

Acknowledgments I gratefully acknowledge Dr. Dagmar Hildebrand and Bianca Kloos for their contributions to the studies on PMT and the review of the manuscript. This work was supported by a grant from the Deutsche Forschungsgemeinschaft (DFG) in the priority program SPP1468-IMMUNOBONE and a Habilitation Fellowship from the Olympia Morata Program of the University of Heidelberg.

References

Adler B, Bulach D, Chung J, Doughty S, Hunt M, Rajakumar K, Serrano M, van Zanden A, Zhang Y, Ruffolo C (1999) J Biotechnol 73:83–90
Akira S, Uematsu S, Takeuchi O (2006) Cell 124:783–801
Aktories K (2011) Nat Rev Microbiol 9:487–498
Amigot JA, Torremorell M, Pijoan C (1998) J Vet Diagn Invest 10:169–173
Anders HJ, Banas B, Schlondorff D (2004) J Am Soc Nephrol 15:854–867
Arron JR, Choi Y (2000) Nature 408:535–536
Arumugam ND, Ajam N, Blackall PJ, Asiah NM, Ramlan M, Maria J, Yuslan S, Thong KL (2011) Trop Biomed 28:55–63
Bagley KC, Abdelwahab SF, Tuskan RG, Fouts TR, Lewis GK (2002a) Infect Immun 70:5533–5539
Bagley KC, Abdelwahab SF, Tuskan RG, Fouts TR, Lewis GK (2002b) J Leukoc Biol 72:962–969
Bagley KC, Abdelwahab SF, Tuskan RG, Lewis GK (2005) Infect Immun 73:413–421
Biberstein EL (1990) Can J Vet Res 54(Suppl):S78–S82
Blander JM, Medzhitov R (2004) Science 304:1014–1018
Blin-Wakkach C, Wakkach A, Quincey D, Carle GF (2006) Exp Hematol 34:851–859
Blocker D, Berod L, Fluhr JW, Orth J, Idzko M, Aktories K, Norgauer J (2006) Int Immunol 18:459–464
Boyce JD, Cullen PA, Nguyen V, Wilkie I, Adler B (2006) Proteomics 6:870–880
Brown J, Wang H, Hajishengallis GN, Martin M (2011) J Dent Res 90:417–427
Broz P, Monack DM (2011) Immunol Rev 243:174–190
Calvani N, Silvestris F, Cafforio P, Dammacco F (2004) Leuk Lymphoma 45:377–380
Carter GR (1955) Am J Vet Res 16:481–484
Chevalier G, Le Henaff M, Wroblewski H (1992) C R Acad Sci III(314):253–258
Chrisp CE, Foged NT (1991) Am J Vet Res 52:56–61
Dabo SM, Taylor JD, Confer AW (2007) Anim Health Res Rev 8:129–150
Dominick MA, Rimler RB (1986) Am J Vet Res 47:1532–1536
Doughty SW, Ruffolo CG, Adler B (2000) Vet Microbiol 72:79–90
Dudet LI, Chailler P, Dubreuil JD, Martineau-Doize B (1996) J Cell Physiol 168:173–182
Faries MB, Bedrosian I, Xu S, Koski G, Roros JG, Moise MA, Nguyen HQ, Engels FH, Cohen PA, Czerniecki BJ (2001) Blood 98:2489–2497
Felix R, Fleisch H, Frandsen PL (1992) Infect Immun 60:4984–4988
Fernandez de Henestrosa AR, Badiola I, Saco M, Perez de Rozas AM, Campoy S, Barbe J (1997) FEMS Microbiol Lett 154:311–316
Finlay BB, McFadden G (2006) Cell 124:767–782
Foged NT (1992) APMIS Suppl 25:1–56
Frehel C, de Chastellier C, Lang T, Rastogi N (1986) Infect Immun 52:252–262
Gagliardi MC, Sallusto F, Marinaro M, Langenkamp A, Lanzavecchia A, De Magistris MT (2000) Eur J Immunol 30:2394–2403
Galdiero M, Folgore A, Nuzzo I, Galdiero E (2000) Immunobiology 202:226–238
Galdiero M, Palomba E, De L, Vitiello M, Pagnini P (1998) Am J Vet Res 59:1270–1274
Gross O, Thomas CJ, Guarda G, Tschopp J (2011) Immunol Rev 243:136–151
Guermonprez P, Valladeau J, Zitvogel L, Thery C, Amigorena S (2002) Annu Rev Immunol 20:621–667
Harper M, Boyce JD, Adler B (2006) FEMS Microbiol Lett 265:1–10
Harper M, Boyce JD, Cox AD, St Michael F, Wilkie IW, Blackall PJ, Adler B (2007) Infect Immun 75:3885–3893
Harper M, Cox AD, Adler B, Boyce JD (2011) Vet Microbiol 153:109–115
Harper M, Cox AD, St Michael F, Wilkie IW, Boyce JD, Adler B (2004) Infect Immun 72:3436–3443

Heddleston KL, Goodson T, Leibovitz L, Angstrom CI (1972) Avian Dis 16:729–734
Hildebrand D, Heeg K, Kubatzky KF (2011) Infect Immun 79:220–228
Hildebrand D, Walker P, Dalpke A, Heeg K, Kubatzky KF (2010) Cell Microbiol 12:1732–1745
Horowitz MC, Bothwell AL, Hesslein DG, Pflugh DL, Schatz DG (2005) Immunol Rev 208:141–153
Hoskins IC, Thomas LH, Lax AJ (1997) Vet Rec 140:22
iDali C, Foged NT, Frandsen PL, Nielsen MH, Elling F (1991) J Gen Microbiol 137:1067–1071
Iovane G, Pagnini P, Galdiero M, Cipollaro de l'Ero G, Vitiello M, D'Isanto M, Marcatili A (1998) Vet Immunol Immunopathol 66:391–404
Jutras I, Martineau-Doize B (1996) Can J Vet Res 60:34–39
Kubin M, Chow JM, Trinchieri G (1994) Blood 83:1847–1855
Lax AJ, Thomas W (2002) Trends Microbiol 10:293–299
Lu YS, Aguila HN, Lai WC, Pakes SP (1991) Infect Immun 59:1470–1475
Luo Y, Zeng Q, Glisson JR, Jackwood MW, Cheng IH, Wang C (1999) Vaccine 17:821–831
Marcatili A, D'Isanto M, Galdiero M, Pagnini U, Palomba E, Vitiello M, Martone F (2000) Res Microbiol 151:217–228
Marques R, Boneca IG (2011) Cell Mol Life Sci 68:3661–3673
Martineau-Doize B, Caya I, Gagne S, Jutras I, Dumas G (1993) J Comp Pathol 108:81–91
May BJ, Zhang Q, Li LL, Paustian ML, Whittam TS, Kapur V (2001) Proc Natl Acad Sci U S A 98:3460–3465
Medzhitov R, Preston-Hurlburt P, Janeway CA Jr (1997) Nature 388:394–397
Mensah KA, Li J, Schwarz EM (2009) Immunol Res 45(2–3):100–113
Miao EA, Rajan JV, Aderem A (2011) Immunol Rev 243:206–214
Michael GB, Kadlec K, Sweeney MT, Brzuszkiewicz E, Liesegang H, Daniel R, Murray RW, Watts JL, Schwarz S (2012) J Antimicrob Chemother 67:91–100
Miyaura C, Onoe Y, Inada M, Maki K, Ikuta K, Ito M, Suda T (1997) Proc Natl Acad Sci U S A 94:9360–9365
Mossman KL, Mian MF, Lauzon NM, Gyles CL, Lichty B, Mackenzie R, Gill N, Ashkar AA (2008) J Immunol 181:6702–6706
Mullan PB, Lax AJ (1998) Calcif Tissue Int 63:340–345
Muller WA (2011) Sci Signal 4:23
Nair SP, Meghji S, Wilson M, Reddi K, White P, Henderson B (1996) Infect Immun 64:2371–2380
Nielsen JP, Foged NT, Sorensen V, Barfod K, Bording A, Petersen SK (1991) Can J Vet Res 55:128–138
Nobes C, Marsh M (2000) Curr Biol 10:R739–R741
Orth JH, Aktories K, Kubatzky KF (2007) J Biol Chem 282:3050–3057
Orth JH, Fester I, Preuss I, Agnoletto L, Wilson BA, Aktories K (2008) J Biol Chem 283:23288–23294
Orth JH, Lang S, Taniguchi M, Aktories K (2005) J Biol Chem 280:36701–36707
Orth JH, Preuss I, Fester I, Schlosser A, Wilson BA, Aktories K (2009) Proc Natl Acad Sci U S A 106:7179–7184
Oswald E, Nougayrede JP, Taieb F, Sugai M (2005) Curr Opin Microbiol 8:83–91
Pettit RK, Ackermann MR, Rimler RB (1993) Lab Invest 69:94–100
Preuss I, Hildebrand D, Orth JH, Aktories K, Kubatzky KF (2010) Cell Microbiol 12:1174–1185
Proft T, Baker EN (2009) Cell Mol Life Sci 66:613–635
Pulendran B, Ahmed R (2011) Nat Immunol 12:509–517
Pullinger GD, Bevir T, Lax AJ (2004) Mol Microbiol 51:255–269
Ramdani and Adler B (1991) Vet Microbiol 26:335–347
Redfield RJ, Findlay WA, Bosse J, Kroll JS, Cameron AD, Nash JH (2006) BMC Evol Biol 6:82
Ribet D, Cossart P (2010) FEBS Lett 584:2748–2758
Richardson M, Fletch A, Delaney K, DeReske M, Wilcox LH, Kinlough-Rathbone RL (1997) Lab Anim Sci 47:27–35
Rimler RB, Phillips M (1986) Avian Dis 30:409–415

Rozengurt E, Higgins T, Chanter N, Lax AJ, Staddon JM (1990) Proc Natl Acad Sci U S A 87:123–127
Ruffolo CG, Tennent JM, Michalski WP, Adler B (1997) Infect Immun 65:339–343
Ryu HI, Kim CJ (2000) J Vet Sci 1:87–95
Sabatte J, Maggini J, Nahmod K, Amaral MM, Martinez D, Salamone G, Ceballos A, Giordano M, Vermeulen M, Geffner J (2007) Cytokine Growth Factor Rev 18:5–17
Shivachandra SB, Viswas KN, Kumar AA (2011) Anim Health Res Rev 12:67–82
Srikumaran S, Kelling CL, Ambagala A (2008) Anim Health Res Rev 8:215–229
Steen JA, Steen JA, Harrison P, Seemann T, Wilkie I, Harper M, Adler B, Boyce JD (2010) PLoS Pathog 6:e1000750
Stuart LM, Ezekowitz RA (2005) Immunity 22:539–550
Townsend KM, Boyce JD, Chung JY, Frost AJ, Adler B (2001) J Clin Microbiol 39:924–929
Underhill DM (2003) Eur J Immunol 33:1767–1775
van der Woude MW, Baumler AJ (2004) Clin Microbiol Rev 17:581–611
van Diemen PM, de Vries Reilingh G, Parmentier HK (1994) Vet Immunol Immunopathol 41:307–321
Vasfi Marandi M, Mittal KR (1997) Infect Immun 65:4502–4508
Wetzler LM (2010) Future Microbiol 5:749–758
Wijewardana TG, Sutherland AD (1990) Vet Microbiol 24:55–62
Wilson BA, Ho M (2011) FEBS J 278:4616–4632
Wilson BA, Zhu X, Ho M, Lu L (1997) J Biol Chem 272:1268–1275
Zidek Z (1999) Eur Cytokine Netw 10:319–328
Zurlo JZ (2005) Pasteurella species. In: Mandell L, Bennett JE, Dolin R (eds) Mandell, Douglas and Bennett's principles and practice of infectious diseases. Elsevier Churchill Livingstone, Philadelphia
Zywietz A, Gohla A, Schmelz M, Schultz G, Offermanns S (2001) J Biol Chem 276:3840–3845

Molecular Biology of *Pasteurella multocida* Toxin

Joachim H. C. Orth and Klaus Aktories

Abstract *Pasteurella multocida* toxin (PMT) is the causative agent of progressive atrophic rhinitis in swine. The 146 kDa single-chain toxin harbours discrete domains important for receptor binding, internalisation and biological activity. The molecular basis of the toxin's activity is the deamidation of a specific glutamine residue in the α-subunit of heterotrimeric G proteins. This results in an inhibition of the inherent GTPase activity leading to a constitutively active phenotype of the G protein. Due to the ability of the toxin to act on various families of heterotrimeric G proteins, a large subset of signal transduction pathways is stimulated.

Contents

J. H. C. Orth (✉) · K. Aktories
Institut für Experimentelle und Klinische Pharmakologie und Toxikologie,
Albert-Ludwigs-Universität Freiburg, Albertstr. 25, 79104 Freiburg, Germany
e-mail: joachim.orth@pharmakol.uni-freiburg.de

K. Aktories
e-mail: klaus.aktories@pharmakol.uni-freiburg.de

Current Topics in Microbiology and Immunology (2012) 361: 73–92
DOI: 10.1007/82_2012_201

Published Online: 28 February 2012

Pasteurella multocida is an opportunistic pathogenic bacterium living in the nasal pharyngeal space of animals. Infections of humans normally arise from scratches and bites by domesticated animals such as cats and dogs. Under special conditions infection of pigs with *P. multocida* leads to an atrophic rhinitis, which is characterised by the atrophy of nasal turbinate bones accompanied by a shortening and twisting of the snout. The causative agent of the atrophic rhinitis was found to be the bacterial protein toxin PMT. The toxin is produced by two (A and D) of five serogroups of *P. multocida* (Frandsen et al. 1991; Williamson 1994).

After entering the cell, the 146 kDa toxin activates various signal transduction pathways by stimulating heterotrimeric G proteins of the $G\alpha_{q/11}$, $G\alpha_{12/13}$ and $G\alpha_i$ family. After a discussion of the structure of PMT, the mechanisms by which PMT enters eukaryotic cells and produces biological effects will be described in detail.

1 Structure of PMT

Similar to many bacterial protein toxins, *Pasteurella multocida* toxin (PMT) is a multifunctional protein comprising different domains with specific functions. The toxin contains a receptor binding and translocation domain (B) and a biologically active (A) domain. Therefore, it can be described as a typical AB-type protein toxin (Fig. 1).

PMT consists of 1285 amino acid residues resulting in a mass of 146 kDa (Buys et al. 1990; Petersen 1990). The receptor binding domain is located in the N-terminal part of the protein, including the amino acid residues 1–580. Within this domain a putative translocation domain was found between residues 402–457. The N-terminus of PMT has homology (~20%) to another family of bacterial protein toxins named cytotoxic necrotizing factors found in uropathogenic *E. coli* (CNF1, CNF2, CNF3) and *Yersinia pseudotuberculosis* (CNFY) (Lemichez et al. 1997; Pullinger et al. 2001). CNFs are deamidases, which act on small GTPases like Rho, Rac or Cdc42 (Flatau et al. 1997; Schmidt et al. 1997). Consistent with this homology, the receptor binding domains of CNF and PMT are located in the N-terminal part and the active domains in the C-terminus (Busch et al. 2001; Pullinger et al. 2001).

The crystal structure of the C-terminal part of PMT revealed three domains designated C1, C2 and C3 (PDB ID 2EBF). The C3 domain is of major importance because it harbours the catalytic activity of the toxin to modify intracellular targets. The overall structure of the crystallised portion was described as Trojan horse-like shape with feet (C1), a body (C2) and a head (C3) (Kitadokoro et al. 2007; Miyazawa et al. 2006).

The feet C1-domain, encompassing amino acid residues 575–719, consists of seven helices. The tertiary structure of the first four helices shows similarity to the N-terminal portion of *Clostridium difficile* toxin B (25% identity) (Kamitani et al. 2010). In toxin B, these four helices serve as a plasma membrane targeting signal. An ectopically expressed GFP-fusion of a C1 fragment of PMT (amino acid

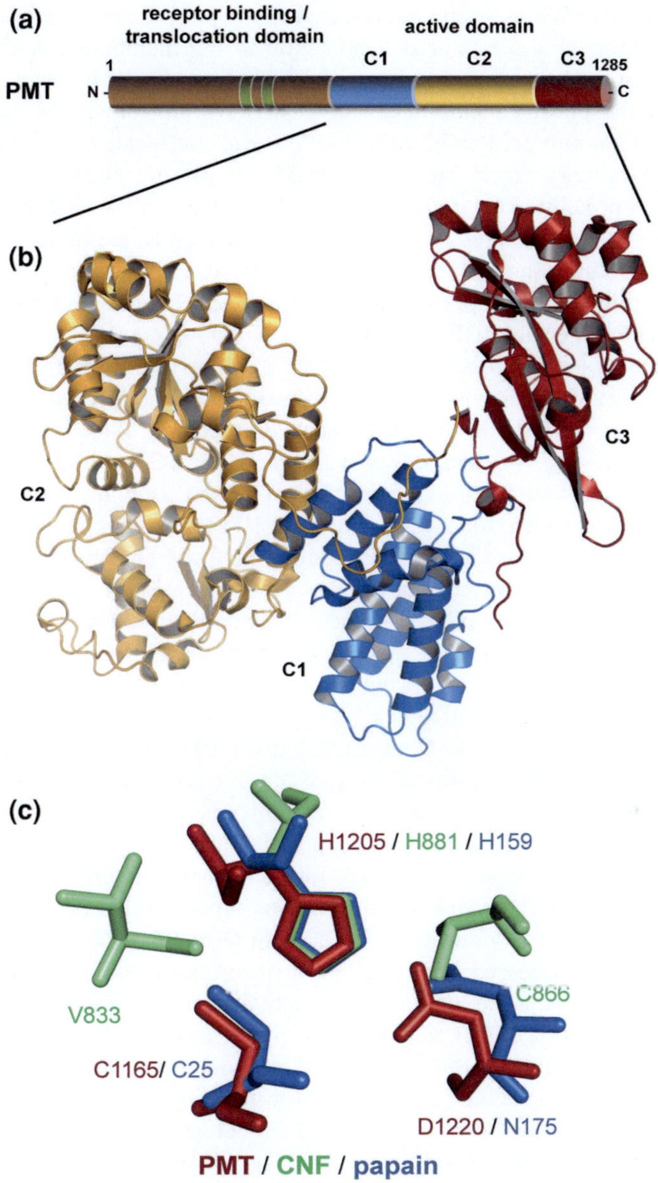

Fig. 1 Overall domain structure of PMT. **a** The toxin consists of an active domain in the C-terminus and a receptor binding and translocation domain in the N-terminus. **b** The crystal structure of a C-terminal fragment shows three distinct domains C1, C2 and C3. C1 is involved in plasma membrane binding. The function of C2 is still unknown and C3 is a deamidase to activate G protein α-subunits. **c** Comparison of the active sites of PMT, CNF1 and papain. Images were generated using PyMol and PDB data files 2EBF (C-PMT), 2EC5 (C-PMT^{C1159S}), 1HQO (CNF1) and 1POP (papain)

residues 569–671) shows strong localisation to the plasma membrane. Deletion of each of the four helices impaired localisation to the membrane. Additionally, a GFP-fusion of a C-terminal fragment of PMT (amino acid residues 569–1285) localised to the plasma membrane, whereas a deletion mutant of the first four helices of C1 (amino acid residues 671–1285) did not localise in the membrane. Because the primary target proteins of PMT are plasma membrane-bound heterotrimeric G proteins, impaired localisation to the substrate would diminish PMT toxicity. Congruently, expression of PMT fragments which are impaired in plasma membrane binding, showed no biological activity. Interestingly, an N-myristoylation peptide tag compensated for the deleted helices and restored PMT activity (Kamitani et al. 2010).

Moreover, this domain was identified as a conserved membrane localisation domain in clostridial glucosyltransferase toxins from *C. difficile*, *C. novyi*, *C. sordellii* or *C. perfringens* and the Multifunctional Autoprocessing RTX toxins (MARTX) from *Vibrio cholerae*, *V. vulnificus* or *V. anguillarum* (Geissler et al. 2009). GFP fusion proteins of these homologue domains are also membrane associated. Three amino acid residues were identified, which are 100% identical between membrane localisation domains. For PMT, these residues are Tyr-611, Ser-651 and Arg-653. Consequences of site-directed mutagenesis of these residues were only tested in MARTX-derived domains, but not in the correspondent PMT domain. The results showed that only the Ser and the Arg are essential for appropriate membrane targeting. Their function could be the maintenance of the overall structure of the four-helix bundle (Geissler et al. 2010).

The largest domain in the C-terminal part of PMT is the so-called body or C2 domain (amino acid residues 720–1,104). C2 consists of 18 helices and nine β-strands and can be divided into two subdomains. Both subdomains exhibit a structure typical of nucleotide-binding proteins. Folylpolyglutamate synthetase and cdc14bs show structural homology to the second subdomain. Because both enzymes interact with a phosphate group it was supposed that this could be a hint to the still unknown function of the C2 domain (Kitadokoro et al. 2007).

Intracellularly expressed PMT and toxin truncations confirmed that the biologically active domain, i.e. the G protein-activating domain is located within C3 (Aminova et al. 2008). This C3 domain (amino acid residues 1,105–1,285) is connected by a long loop (1,087–1,104) to the C2 domain. C3 is separated into two subdomains and provides the catalytic cleft for the enzymatic function of the toxin. Interestingly, a disulfide bond was found in C3 between Cys-1159 and Cys-1165. Mutational studies of these residues revealed that the toxic activity of PMT strictly depends on Cys-1165 but not Cys-1159 (Busch et al. 2001; Kitadokoro et al. 2007; Ward et al. 1998). In addition, His-1205, His-1223, Asp-1220 and Gln-1225 are essential for PMT activity (Kitadokoro et al. 2007; Orth et al. 2003; Pullinger and Lax 2007). Structural analysis of PMT mutants in which the disulfide bonds were ablated by replacing the respective Cys-1159 or Cys-1165 residues by Ser (PDB ID 2EC5/2EBH) displayed different structures compared to wt-PMT (PDB ID 2EBF). Most striking is a reorientation of Cys-1165 when Cys-1159 is replaced by Ser. Cys-1165 is displaced towards the catalytic cleft of the C3 domain and forms a thiolate imidazolium ion pair with His-1205. The secondary structure of the reoriented C3 domain of $PMT^{Cys1195Ser}$ shows homology to the cysteine protease AvrPphB from *Pseudomonas syringae*, which in turn has structural similarity to the papain-like cysteine protease fold (Kitadokoro et al. 2007). The catalytic triad of PMT consists, therefore, of Cys-1165, His-1205 and Asp-1220 (Kitadokoro et al. 2007). The structure of the catalytic triad is very similar to the catalytic triads of transglutaminases like Factor XIII (PDB ID 1GGT) or fish derived transglutaminases (PDB ID 1GOD), as shown by superimposing the catalytic His residues. Additionally, PMT and papain exhibit similar topography of their catalytic triads (Fig. 1) (Orth et al. 2009). Despite the fact that PMT is a deamidase but not a transglutaminase nor thiol protease, it is not contradictory that the toxin shares the topography of the active sites of these enzymes. Transglutaminases replace the NH_2-group of the amide of glutamine by another amine group. Deamidases catalyse a similar type of chemical reaction, converting the amide into a carboxylate utilising water instead of an amine. Finally, transglutaminases perform the inverse reaction to thiol proteases. Therefore, the biological activity of PMT is supported by the crystal structure of the C3 domain.

The contribution of His-1223 and Gln-1225, which are also essential for toxin activity, is not completely understood to date. Both amino acid residues are in close proximity to the catalytic triad, and at least for Gln-1225, a direct participation in enzymatic activity was proposed (Kitadokoro et al. 2007).

2 Uptake into Eukaryotic Cells

Knowledge of the uptake of PMT into eukaryotic cells is limited. Three major steps are generally involved in the entry of bacterial protein toxins into their target cells. (i) Binding of a cell membrane receptor. (ii) Internalisation by endocytosis. (iii) Release of the biologically active component or the complete toxin into the cytosol (Sandvig and van Deurs 2005). For receptor binding and translocation, the N-terminal part of PMT is responsible (amino acids 1–580). Whereas the translocation domain is narrowed down to amino acid residues 402–457, the receptor binding region is not specified (Baldwin et al. 2004).

The PMT receptor is not yet characterised. Previous studies with colloidal, gold-labelled toxin observed binding to Vero cells. Incubation of mixed gangliosides with PMT inhibited toxin binding to Vero cells, indicating a competition between the added gangliosides and the PMT receptor. Therefore, it was suggested that PMT interacts with a ganglioside-type receptor (Pettit et al. 1993). More recently, these results were challenged by new findings. Using different approaches such as TLC-overlay, surface plasmon resonance analysis or toxin pull-down of liposomes, binding of PMT to membrane phospholipids like sphingomyelin and phosphatidylcholine was detected (Brothers et al. 2011). In contrast to the previous study, no binding of PMT to gangliosides like $GM_{1/2/3}$ was found. Interestingly, an interaction of a C-terminal fragment of PMT encompassing the crystallised domains C1, C2 and C3 with GM_1 was discovered. The authors suggested that removal of the PMT N-terminus unmasks the membrane binding site of the C1 domain.

Surface plasmon resonance analysis leads to an at least bi-phasic binding of PMT to cells. After initial binding with low affinity to an abundant membrane component, a more specific binding to sphingomyelin could follow. The involvement of an additional proteinaceous receptor, which would induce endocytosis, was proposed (Brothers et al. 2011).

After binding to the receptor at the plasma membrane, the toxin enters the cell by endocytosis. Utilising various inhibitors of vesicle trafficking and GFP-fusions of PMT fragments, the intracellular pathway of endocytosed toxin was followed. GFP-fusions of the N-terminal portion of PMT harbouring the putative receptor binding and translocation domain were found to colocalize with transferrin receptor and at early time points with cholera toxin B subunit, indicating localisation in early endosomes (Repella et al. 2011). After passing early endosomes, transferrin receptor traffics to recycling endosomes and cholera toxin via the Golgi apparatus to the endoplasmic reticulum. Conversely, PMT is supposed to translocate to late endosomes where the translocation to the cytosol occurs (Repella et al. 2011). In line with this model, inhibitors of trafficking between Golgi apparatus and endoplasmic reticulum, like brefeldin A, do not inhibit the biological activity of PMT. However, cell entry depends on Arf6. The small GTPase Arf6 binds to endosomes and is important for the trafficking of recycling endosomes (Peters et al. 1995). Both a dominant-negative form and a constitutively cycling mutant of Arf6 inhibited PMT intoxication, implicating the involvement of this GTPase in toxin uptake (Repella et al. 2011).

It is thought that PMT translocates on the way from early to late endosomes to the cytosol. The acidification of the endosome plays a pivotal role for translocation (Baldwin et al. 2004; Rozengurt et al. 1990). Consequently, inhibition of endosome acidification by blockade of the vacuolar H^+ ATPases using bafilomycin A1 represses toxin activity (Baldwin et al. 2004). In addition, the direct transfer of plasma membrane bound toxin to the cytosol is inducible by mimicking the endocytic conditions, e.g. applying acidic medium to cells (Baldwin et al. 2004). The translocation depends on a putative translocation T-domain, consisting of two predicted hydrophobic helices (residues 402–423 and 437–457) linked by a peptide loop (residues 424–436). It is proposed that acidification induces a structural change in the toxin, which was previously observed utilising circular dichroism and measuring susceptibility to proteases (Smyth et al. 1995, 1999). This structural change exposes the T-domain, allowing it to insert into the vesicular membrane. Mutational studies of amino acid residues in the peptide loop between the hydrophobic helices suggest that acidic residues in this region are of major importance for membrane insertion (Baldwin et al. 2004).

So far, our knowledge of the membrane translocation process itself is scant. It is suggested that the toxin at least partly unfolds. Whether the toxin spontaneously refolds in the cytosol or whether chaperons support this process is not known.

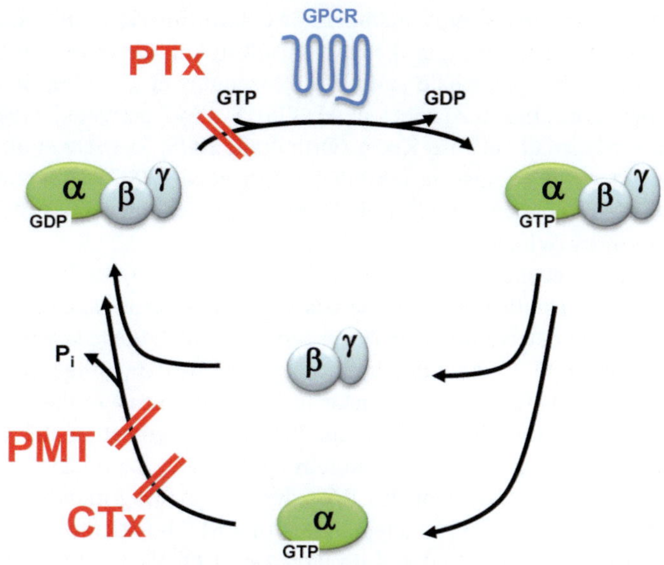

Fig.2 GTPase cycle of heterotrimeric G proteins. Inhibition of the GTPase cycle by PMT, cholera toxin (CTx) and pertussis toxin (PTx) at specific steps within the cycle is indicated

3 Activation of Heterotrimeric G Proteins

PMT activates signal transduction pathways, which are downstream of heterotrimeric G proteins. Heterotrimeric G proteins are molecular switches cycling between a GTP- and GDP-bound state (Bourne et al. 1990; Gilman 1987; Oldham and Hamm 2006, 2008; Sprang 1997). The inactive heterotrimeric complex of the α- and the non-dissociable βγ-subunit binds GDP. An agonist-activated heptahelical receptor, also known as G protein-coupled receptor (GPCR), functions as a guanine nucleotide exchange factor (GEF). The activated GPCR induces conformational changes in the α-subunit to release the GDP and to rebind GTP (Gilman 1987; Rasmussen et al. 2011; Westfield et al. 2011). After binding GTP, the heterotrimeric complex dissociates into the GTP-bound α-subunit and the βγ-subunit. Both can interact and stimulate their own effectors. The inherent GTPase activity of the α-subunit, which can be stimulated by regulators of G protein signalling (RGS) (Hollinger and Hepler 2002; Ross and Wilkie 2000), hydrolyses the γ-phosphate of GTP. The GDP-bound α-subunit reassociates with Gβγ and signal transduction is terminated (Fig. 2). The heterotrimeric G proteins are grouped with respect to the effector interactions of the α-subunits: The G_s family activates the adenylyl cyclase (Northup et al. 1980), $G_{i/o}$ proteins inhibit the adenylyl cyclase (Bokoch et al. 1983), $G_{q/11}$ stimulates the phospholipase C (PLC) β1 (Taylor et al. 1990) and $G_{12/13}$ activates the small GTPase Rho via RhoGEF proteins (Birnbaumer 2007a, b; Hepler and Gilman 1992; Riobo and Manning 2005).

PMT activates diverse family members of heterotrimeric G proteins. Via $G\alpha_q$ the toxin leads to activation of PLCβ1, resulting in increased levels of diacylglycerol, inositoltrisphosphate and Ca^{2+} (Rozengurt et al. 1990; Staddon et al. 1991; Wilson et al. 1997; Zywietz et al. 2001). PMT-activated $G\alpha_{13}$ leads to stimulation of the small GTPase RhoA (Orth et al. 2005; Zywietz et al. 2001) and via $G\alpha_i$ the adenylyl cyclase is inhibited (Orth et al. 2008). A comprehensive overview of PMT-induced signalling is lined out in Cellular effects of Pasteurella multocida toxin by Wilson.

The molecular mechanism by which PMT activates signalling via heterotrimeric G proteins was elucidated on the basis of $G\alpha_i$. One advantage of $G\alpha_i$ is the possibility to easily determine the GTPase activity, which is the key in terminating G protein signalling. A so-called multiple cycle GTPase assay (Aktories and Jakobs 1981) in membrane preparations was utilised to measure the effect of PMT on heterotrimeric G proteins (Orth et al. 2008). An agonist of the G_i-coupling EDG-receptor was used to induce G protein cycling and the released γ-phosphate was measured. The receptor agonist LPA exhibited a strong induction of GTPase activity of $G\alpha_i$. Interestingly, in membranes of PMT-intoxicated cells the basal GTPase activity was diminished and no increase of GTPase activity was induced by receptor agonists (Orth et al. 2008). The inhibition of GTPase activity induced by PMT suggested that the toxin activated G protein signalling by blocking the terminating GTP hydrolysis. However, the multiple cycle GTPase assay was not appropriate to determine unequivocally the step of GTP hydrolysis. Besides inhibition of the GTPase activity, the toxin caused uncoupling of $G\alpha_i$ from its receptor. This was measured by receptor-induced GTPγS-binding, which was blocked by PMT treatment (Orth et al. 2007).

The effect of PMT was compared to that of pertussis toxin (PTx). PTx ADP-ribosylates $G\alpha_i$ proteins and inhibits the interaction of the heterotrimeric G protein with the GPCR (Gierschik 1992; Katada and Ui 1982; Murayama and Ui 1983; Nürnberg 1997; Ui 1984). Thereby the activation and the cycling of the G protein are blocked. The observed outcome is the inhibition of the GTPase activity and receptor-induced GTPγS-binding. In respect to the effects measured (e.g., inhibition of steady-state GTP hydrolysis and blockade of agonist-induced GTPγS-binding), PMT and PTx caused similar results, however, PTx blocked $G\alpha_i$ signalling and PMT stimulated $G\alpha_i$ signalling as revealed by inhibition of the adenylyl cyclase. This indicated that PMT disrupts GTPase cycling at a different step as PTx.

A major experimental advantage of using $G\alpha_i$ compared to other G protein α-subunits is the possibility of expressing the functional recombinant protein in high amounts. Therefore, $G\alpha_{i2}$ was coexpressed with the toxin in *E. coli*. Subsequently, the purified G protein was utilised for a single turnover GTPase assay. This kind of assay enables the determination of the GTPase reaction itself and not only the complete cycle (e.g., nucleotide-binding, hydrolysis and release). The intrinsic GTPase activity of $G\alpha_{i2}$, which was coexpressed with the inactive PMT mutant Cys-1165-Ser, was stimulated by addition of regulator of G protein signalling (RGS)3s or RGS16. Both RGS proteins function as GTPase activating proteins

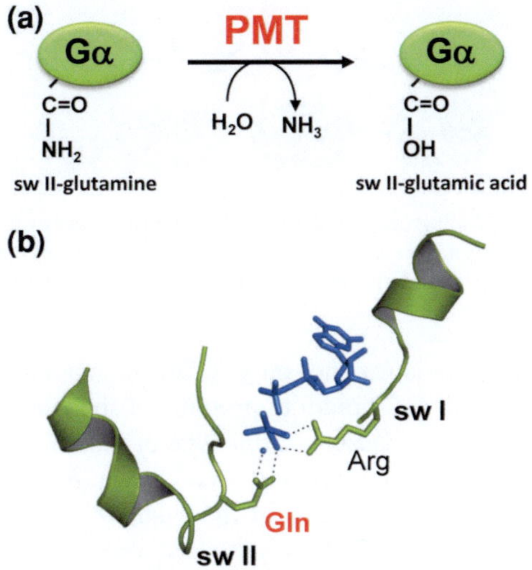

Fig. 3 Molecular mechanism of PMT to activate heterotrimeric G proteins. The toxin deamidates a specific glutamine residue in the α-subunit of heterotrimeric G proteins leading to a glutamic acid residue in this position. **a** Structural overview of the switch (sw) I and II region in $G\alpha_{i1}$ binding to GDP-AlF_4^-, mimicking the transition state complex of the hydrolysis of the γ-phosphate. The conserved Gln and Arg residues participate in the coordination of the γ-phosphate and the attacking water nucleophile (*blue spot*). Image was generated using PyMol and PDB data file 1GFI

(GAP) to facilitate the intrinsic GTPase reaction of the α-subunit. $G\alpha_{i2}$ coexpressed with active PMT showed decreased basal GTP-hydrolysing activity and RGS proteins did not increase this activity over the basal level (Orth et al. 2009). These results defined the hydrolysis of GTP as the critical step inhibited by the toxin's action.

Tandem mass spectrometric analysis of recombinant $G\alpha_{i2}$, which was coexpressed with PMT, revealed a 1 Da change in the switch II region of the α-subunit. A peptide of the switch II region, encompassing amino acid residues 199-MFDVGG**Q**R-206, was recovered after tryptic digestion and MS analysis of $G\alpha_{i2}$ coexpressed with inactive toxin. The corresponding peptide of $G\alpha_{i2}$ coexpressed with active toxin was also found. Additionally, a second peptide with a 1 Da higher mass appeared. Tandem MS analysis revealed the following peptide: 199-MFDVGG**E**R-206. Hence, the toxin deamidates Gln-205 of $G\alpha_{i2}$ resulting in glutamic acid in this position (Fig. 3).

The glutamine residue in the switch II region is conserved throughout all α-subunits of heterotrimeric G proteins. This Gln stabilises the pentavalent transition state of the GTP hydrolysis and coordinates the water nucleophile attacking the γ-phosphate of the nucleotide (Coleman et al. 1994; Sprang 1997; Tesmer et al. 1997). Mutations of this site are commonly known to inhibit the GTPase activity consequently leading to a constitutive activation of G protein signalling

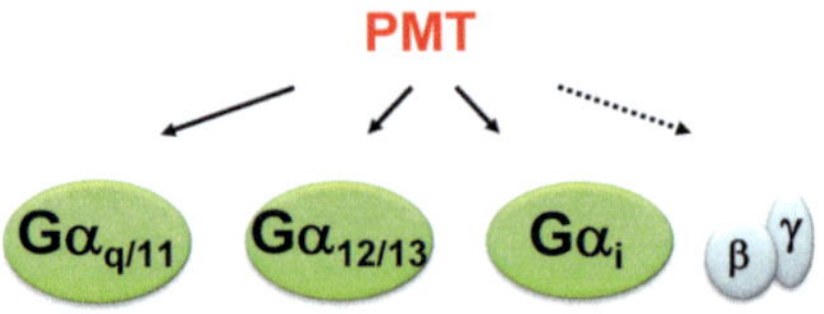

Fig. 4 Heterotrimeric G proteins activated by PMT. The toxin acts on the $G_{q/11}$-, $G_{12/13}$- and G_i-family to induce their signal transduction via their appropriate effectors. By activating the α-subunit the corresponding βγ-subunit is released and stimulates its own signalling cascade

(De Vivo et al. 1992). This is true not only for the α-subunits of heterotrimeric G proteins, but also for small GTP-binding proteins of the Ras superfamily (Bourne et al. 1989). Another important aspect of this type of mutation of G proteins is their frequent finding in mammalian tumours. As known for the transforming Ras mutation at position Gln-61, mutations of the α-subunit of heterotrimeric G proteins, causing inhibition of the GTPase activity, were described as transforming oncogenes and were found in diverse types of tumours (Kalinec et al. 1992; Radhika and Dhanasekaran 2001). For example, mutations of $G\alpha_{i2}$ at position Gln-205 have been observed in pituitary adenomas and mutations of Gln-209 of $G\alpha_q$ were found in melanoma of the uvea and blue naevi (Van Raamsdonk et al. 2008; Williamson et al. 1995).

To further verify that an exchange of the conserved Gln to Glu, as catalysed by PMT, blocks GTP hydrolysis, the mutation was introduced in $G\alpha_{i2}$ (Gln-205) and $G\alpha_q$ (Gln-209). Both mutations constitutively activated the G proteins leading to an inhibition of adenylyl cyclase or stimulation of PLCβ1, respectively (Orth et al. 2009). All these data show that PMT deamidates an essential Gln residue in the switch II region of the α-subunit of heterotrimeric G proteins. The resulting Glu residue is not capable of hydrolysing the bound nucleotide, leading to a constitutive active phenotype of the G protein.

Another amino acid residue contributing to the γ-phosphate coordination of GTP is the Arg in the switch I region. Interestingly, this Arg in the switch I region of $G\alpha_s$ is the target site for ADP-ribosylation by cholera toxin (Dop Van et al. 1984; Freissmuth and Gilman 1989), a key feature of cholera pathogenesis. Accordingly, the GTPase activity of CTx-modified $G\alpha_s$ is blocked and the adenylyl cyclase is stimulated by constitutively active $G\alpha_s$ (Cassel and Selinger 1977).

4 Substrate Specificity

As mentioned above, the PMT-targeted Gln in the switch II region of heterotrimeric G protein α-subunits is conserved throughout all members of G proteins. However, the toxin-induced activation has been verified for only a subset of heterotrimeric G proteins. Initially, the activation of heterotrimeric G proteins by PMT was studied indirectly by measurement of the specific downstream signalling of the respective G proteins (Higgins et al. 1992; Rozengurt et al. 1990; Staddon et al. 1990, 1991). By different methods, members of the $G\alpha_q$-, $G\alpha_{13}$- and $G\alpha_i$-family have been identified as substrates of the toxin. Identification of the primary molecular mode of action of PMT as a deamidation reaction allows the direct study of the covalent modification of toxin-targeted G proteins (Fig. 4).

Most importantly, the deamidation can be verified by mass spectrometric analysis of recombinant protein for $G\alpha_{i2}$. Moreover, the shift of the isoelectric point (pI), resulting from the deamidation of the Gln to the more acidic Glu, is detectable by 2D-gel electrophoresis. For example, a change of 0.07 units was calculated and verified for $G\alpha_{i2}$ and $G\alpha_{i1}$, which were recovered from PMT-treated cells (Orth et al. 2009).

$G\alpha_q$ is a well-known target of PMT (Wilson et al. 1997). Several biochemical data support this view. First, as mentioned above, native gel electrophoresis demonstrates a shift of $G\alpha_q$ after PMT intoxication, indicating a change of the pI induced by deamidation (Orth et al. 2009). More directly, deamidation of $G\alpha_q$ was verified by a monoclonal antibody, which specifically detects the switch II region of $G\alpha_q$ covering the deamidated Gln-209 (i.e. Glu-209) (Kamitani et al. 2011). Because the monoclonal antibody discriminates perfectly between unaffected G proteins and PMT-deamidated G proteins, it is a useful tool to verify the toxin's action at the G protein level. This antibody was employed to answer the puzzle of PMT substrate specificity, which mystified studies with the toxin for years. Thus, several previous studies using mouse embryonic fibroblasts deficient for $G\alpha_q$, $G\alpha_{11}$ or both, showed that PMT-induced $G\alpha_{q/11}$-dependent signalling to the PLCβ1 exclusively via $G\alpha_q$ but not via $G\alpha_{11}$ (Orth et al. 2004; Zywietz et al. 2001). These findings suggested that $G\alpha_{11}$ was not modified by PMT, although the switch II region shares high homology between all heterotrimeric G proteins and is even identical between $G\alpha_q$ and $G\alpha_{11}$ (Orth et al. 2004; Zywietz et al. 2001). However, the new findings obtained by deamidation-specific antibody and now confirmed by MS analysis in our laboratory (J.H.C. Orth and K. Aktories, unpublished data) indicate that both $G\alpha_q$ and $G\alpha_{11}$ are deamidated by PMT. These results suggest that differences in toxin-activated $G\alpha_{q/11}$-signalling are not based on the toxin-substrate interaction per se, but possibly on the interaction of the deamidated G proteins ($G\alpha_q$ and $G\alpha_{11}$) with their effectors.

As for $G\alpha_{q/11}$, cellular effects or signal transduction events were utilised to demonstrate PMT-induced activation of the $G\alpha_{12/13}$ family. A common effector of $G\alpha_{12/13}$ and $G\alpha_{q/11}$ is the small GTPase RhoA (Vogt et al. 2003). RhoGEF proteins, which can be exclusively activated by $G\alpha_{12/13}$ (p115RhoGEF), $G\alpha_{q/11}$ (p63RhoGEF) or by both (LARG) connect the heterotrimeric G proteins with the small G protein RhoA (Booden et al. 2002; Hart et al. 1998; Kozasa et al. 1998; Lutz et al. 2005). Consequently, PMT treatment of cells leads to RhoA-dependent reorganisation of the actin cytoskeleton, stress fibre formation and downstream, to an increase in endothelial permeability (Dudet et al. 1996; Essler et al. 1998; Lacerda et al. 1996; Zywietz et al. 2001). Utilising mouse embryonic fibroblasts deficient for $G\alpha_{q/11}$ or $G\alpha_{12/13}$, the involvement of $G\alpha_{12/13}$ besides $G\alpha_{q/11}$ in PMT-induced RhoA activation was demonstrated (Zywietz et al. 2001). Another method of differentiating between $G\alpha_{q/11}$- and $G\alpha_{12/13}$-mediated signalling is the cyclic peptide YM-254890, which is a specific inhibitor of $G\alpha_{q/11}$ signalling (Takasaki et al. 2004). Using both approaches (e.g., genetic knock-out and YM-254890), the dissection of PMT-induced G protein signalling is possible. Most convincingly, the rescue of $G\alpha_{13}$ in $G\alpha_{12/13}$-deficient cells treated with the $G\alpha_{q/11}$ inhibitor reconstitutes toxin-induced RhoA activation (Orth et al. 2005).

Taken together, the activation of heterotrimeric G proteins by PMT is demonstrated for $G\alpha_i$, $G\alpha_q$ and $G\alpha_{13}$ by measuring G protein signalling. Studies on the G protein itself show deamidation of $G\alpha_{i1}$, $G\alpha_{i2}$ and $G\alpha_q$ by a deamidation-triggered change of the pI or $G\alpha_q$ and $G\alpha_{11}$ by recognition by a specific antibody. The most stringent evidence for PMT-induced activation is MS analysis. Unfortunately, up to now this has been described only for $G\alpha_{i2}$.

5 G Protein Prerequisites for PMT Action

5.1 Role of *Gβγ*

Heterotrimeric G proteins consist of the nucleotide-binding α-subunit and the non-dissociable βγ-subunit (Gilman 1987; Oldham et al. 2006; Sprang 1997). After GPCR-induced activation and subsequent dissociation of Gα from Gβγ, both parts induce downstream signalling. Therefore, various questions arise concerning PMT action and Gβγ: (i) Does PMT induce dissociation of the heterotrimeric complex? (ii) Is Gβγ signalling stimulated by PMT? (iii) Does the cycling between heterotrimeric complex and the dissociated form influence the activation by PMT?

Classical studies have shown that the heterotrimeric complex and not the free α-subunit of $G\alpha_i\beta\gamma$ is the preferred substrate of PTx, allowing the determination of the formation of the heterotrimeric complex by PTx-catalysed ADP-ribosylation (Katada et al. 1986). Treatment of mammalian cells with PMT as well as coexpression of $G\alpha_{i2}$ with PMT leads to an inhibition of PTx-induced ADP-ribosylation of $G\alpha_i$. Additionally, the binding of Gβγ to $G\alpha_{i2}$, coexpressed with active toxin, is reduced to the same extent as $G\alpha_{i2}$ is deamidated by PMT ($\sim$50%) (Orth et al. 2008, 2009). Therefore, a dissociation of the heterotrimeric complex by PMT is likely.

PMT-induced activation of Gβγ signalling is most convincingly demonstrated by determination of the activity of PI3Kγ (Preuss et al. 2009). PI3Kγ is an established effector of βγ-subunits leading to PIP_3 formation (Maier et al. 1999; Stephens et al. 1997). Therefore, PMT induces the translocation of a GFP sensor protein (GFP-$Grp1_{PH}$), which interacts with PIP_3, to the plasma membrane (Preuss et al. 2009). Moreover, using scavengers of Gβγ, like GRK2-CT (G-protein-coupled-receptor kinase 2 C-terminus) (Wu et al. 1998), the direct involvement of Gβγ in PMT-induced PI3Kγ activation can be demonstrated.

Sequestration of $\beta\gamma$-subunits not only blocks $G\beta\gamma$ downstream signalling, but also activation of $G\alpha$-dependent signalling by PMT. Overexpression of GRK2-CT or phosducin, both proteins that sequester $G\beta\gamma$ (Hawes et al. 1994; Wu et al. 1998), strongly reduces PMT activation of $G\alpha$-dependent signalling (Preuss et al. 2009). Moreover, two mutants of $G\alpha_q$ (I25A/E26A and G208A) were utilised to study the role of $G\beta\gamma$ in more detail. The mutants manipulate the interaction of $G\alpha$ with $G\beta\gamma$. The double mutant of $G\alpha_q$ (I25A/E26A) does not bind $G\beta\gamma$, while $G\alpha_q^{G208A}$ binds with increased affinity to $G\beta\gamma$ (Jetzt et al. 2003; Lee et al. 1992). Because both mutants inhibit PMT-induced $G\alpha_q$ activation, it is likely that the binding capability of $G\alpha$ to $G\beta\gamma$ itself is not important, but rather the cycling of the heterotrimeric complex, e.g. the association and dissociation of $G\beta\gamma$. Therefore, it has been suggested that during cycling of the G protein a favourable structure serves as target for the toxin (Preuss et al. 2009). Furthermore, Kamitani et al. tested in an in vitro system whether PMT prefers the monomeric $G\alpha$ or the heterotrimeric complex $G\alpha\beta\gamma$ for deamidation (Kamitani et al. 2011). These studies showed that both states are recognised by the toxin, but the monomeric α-subunit is two orders of magnitude less sensitive than the heterotrimeric complex. Thus, the presence of $G\beta\gamma$ at least enhances PMT action. On the other hand, toxin-induced deamidation during coexpression in *E. coli* occurs without $G\beta\gamma$ (Orth et al. 2009), indicating that $G\beta\gamma$ is not absolutely required for modification of the α-subunit by PMT.

5.2 PMT-Induced Activation of G Proteins is Independent of GPCR Interaction

The interaction, i.e. activation of heterotrimeric G proteins by GPCR, depends mainly on the C-terminal amino acid residues of the $G\alpha$ protein. The last five amino acids are essential for interaction and determine the specificity of G protein-GPCR coupling (Conklin et al. 1993; Hamm 1998; Parekh 2006). On the basis of $G\alpha_q$ it was studied whether PMT-induced activation depends on any receptor interaction (Orth et al. 2007). Therefore, G protein constructs, which cannot couple to GPCRs or G protein-receptor fusion proteins, were tested for PMT-induced activation. As expected, the C-terminal deletion mutant of $G\alpha_q$ was not stimulated by a Gq-coupling receptor. In contrast, the toxin was still able to stimulate this deletion mutant, indicating that the function of PMT is independent of the coupling of the G protein with GPCRs. Vice versa, it was tested whether the toxin also activates an α_{1b}-adrenoceptor-$G\alpha_q$ chimera. This chimera was also stimulated by the toxin, supporting the view that PMT acts completely independently of any receptor interaction (Orth et al. 2007).

In agreement with these results, it was shown that PTx does not block PMT-induced $G\alpha_i$ activation (Orth et al. 2008). PTx ADP-ribosylates a Cys residue in the C-terminus of $G\alpha_{i/o}$ proteins and inhibits interaction of the G protein with the

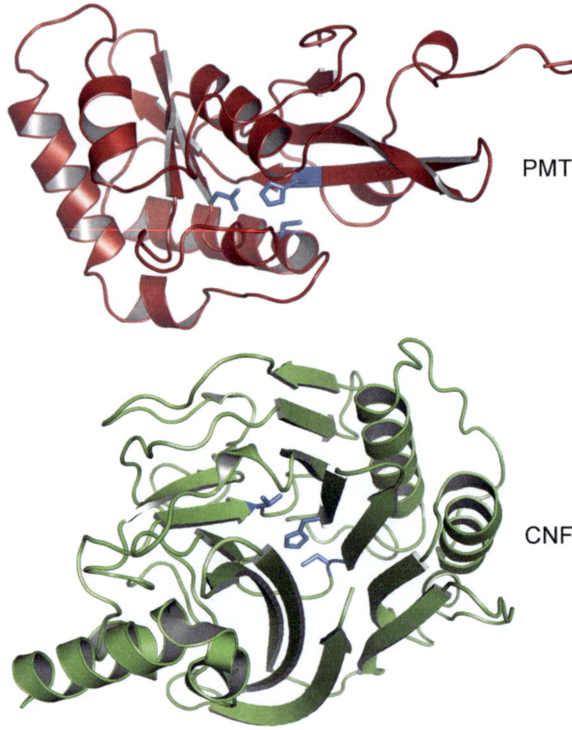

Fig. 5 Comparison of the catalytic domains of PMT and CNF1. Crystal structures of the catalytic active domains of PMT and CNF1. Catalytic triads (*blue sticks*) of PMT and CNF1 are indicated in the folds. Images were generated using PyMol and PDB data files 2EC5 (C-PMT^{C1159S}) and 1HQO (CNF1)

receptor (Katada et al. 1982; Murayama et al. 1983). Even after PTx intoxication, PMT is able to stimulate $G\alpha_i$ leading to an inhibition of the adenylyl cyclase.

6 PMT and Cytotoxic Necrotizing Factors

PMT belongs to a large group of deamidating toxins and/effectors, which cause major pathophysiological alterations of target cells by removing the amide functional group from a specific glutamine residue of the targeted protein substrate. While the targets of PMT are the α-subunits of heterotrimeric G proteins, the cytotoxic necrotizing factors (CNF) 1–3 from *E. coli* and CNFY from *Y. pseudotuberculosis* deamidate small GTPases of the Rho family (Flatau et al. 1997; Hoffmann et al. 2004; Hoffmann and Schmidt 2004; Schmidt et al. 1997; Stoll et al. 2009). It is fascinating that CNFs target the functionally equivalent glutamine residue (e.g., Gln-61 of Rac and Cdc42 or Gln-63 of Rho) of GTPases, resulting in inhibition GTP hydrolysis and constitutive activation of the small G proteins as known for PMT and heterotrimeric G proteins (Aktories 2011).

PMT and CNFs share significant amino acid sequence homology in the N-terminal receptor binding and translocation domain (Lemichez et al. 1997;

Pullinger et al. 2001). However, in contrast to the functionally related eukaryotic targets and type of reaction, no homology is observed in the primary sequence or the tertiary structure of the catalytic active deamidase domain. Moreover, the topography of the catalytic centres of PMT and CNF1 are different (Fig. 5). The catalytic triads of PMT and CNF share the catalytic active Cys and His residues. In the case of PMT, the third catalytic active residue is an Asp and for CNF it is a Val. Superimposing the active His residues revealed that the catalytic triads of PMT and CNF do not match (Fig. 1c). All structural data together indicate that the molecular mechanism of PMT and CNFs to activate G proteins was independently developed during evolution.

7 Outlook

For more than 20 years after the discovery of PMT, its molecular mode of action remained enigmatic. With the elucidation of the toxin as a deamidase, a new era of PMT research has developed. There are still important questions to answer. The precise structural determinants of substrate (G protein) recognition by the toxin are not known. The uptake of the toxin is still not well understood. Are domains 1 and/or domain 2 of the Trojan horse of the C-terminal part of PMT involved in uptake or processing? The answers to these important questions will help to further understand the molecular biology of the toxin. Another important area which has still to be studied is the functional consequence of the actions of the toxin, especially with respect to its pathophysiological role and mechanism in disease. What are the precise pathophysiological steps finally leading to the sequelae of bone destruction of the nose? Thus, we have much to learn about PMT.

Acknowledgments Work of the authors was supported by the German Research Foundation (DFG) through Collaborative Research Centre (SFB) 746 and the Cluster of Excellence 294 BIOSS.

References

Aktories K (2011) Bacterial protein toxins that modify host regulatory GTPases. Nat Rev Microbiol 9:487–498

Aktories K, Jakobs KH (1981) Epinephrine inhibits adenylate cyclase and stimulates a GTPase in human platelet membranes via α-adrenoceptors. FEBS Lett 130(2):235–238

Aminova LR, Luo S, Bannai Y, Ho M, Wilson BA (2008) The C3 domain of Pasteurella multocida toxin is the minimal domain responsible for activation of Gq-dependent calcium and mitogenic signaling. Protein Sci 17:1–5

Baldwin MR, Lakey JH, Lax AJ (2004) Identification and characterization of the Pasteurella multocida toxin translocation domain. Mol Microbiol 54:239–250

Birnbaumer L (2007a) Expansion of signal transduction by G proteins. The second 15 years or so: from 3 to 16 alpha subunits plus betagamma dimers. Biochem Biophys Acta 1768:772–793

Birnbaumer L (2007b) The discovery of signal transduction by G proteins: a personal account and an overview of the initial findings and contributions that led to our present understanding. Biochem Biophys Acta 1768:756–771

Bokoch GM, Katada T, Northup JK, Hewlett EL, Gilman AG (1983) Identification of the predominant substrate for ADP-ribosylation by islet activating protein. J Biol Chem 258:2072–2075

Booden MA, Siderovski DP, Der CJ (2002) Leukemia-associated Rho guanine nucleotide exchange factor promotes G alpha q-coupled activation of RhoA. Mol Cell Biol 22: 4053–4061

Bourne HR, Landis CA, Masters SB (1989) Hydrolysis of GTP by the alpha-chain of Gs and other GTP binding proteins. Proteins 6:222–230

Bourne HR, Sanders DA, McCormick F (1990) The GTPase superfamily: a conserved switch for diverse cell functions. Nature 348:125–132

Brothers MC, Ho M, Maharjan R, Clemons NC, Bannai Y, Waites MA, Faulkner MJ, Kuhlenschmidt TB, Kuhlenschmidt MS, Blanke SR, Rienstra CM, Wilson BA (2011) Membrane interaction of Pasteurella multocida toxin involves sphingomyelin. FEBS J 278(23):4633–4648

Busch C, Orth J, Djouder N, Aktories K (2001) Biological activity of a C-terminal fragment of *Pasteurella multocida* toxin. Infect Immun 69:3628–3634

Buys WEC, Smith HE, Kamps AMIE, Smits MA (1990) Sequence of the dermonecrotic toxin of *Pasteurella multocida* ssp. *multocida*. Nucl Acids Res 18:2815–2816

Cassel D, Selinger Z (1977) Mechanism of adenylate cyclase activation by cholera toxin: inhibition of GTP hydrolysis at the regulatory site. Proc Natl Acad Sci U S A 74:3307–3311

Coleman DE, Berghuis AM, Lee E, Linder ME, Gilman AG, Sprang SR (1994) Structures of active conformations of Gi alpha 1 and the mechanism of GTP hydrolysis. Science 265: 1405–1412

Conklin BR, Farfel Z, Lustig KD, Julius D, Bourne HR (1993) Substitution of three amino acids switches receptor specificity of Gq alpha to that of Gi alpha. Nature 363:274–276

De Vivo M, Chen J, Codina J, Iyengar R (1992) Enhanced phospholipase C stimulation and transformation in NIH-3T3 cells expressing Q209LGq-alpha-subunits. J Biol Chem 267: 18263–18266

Dop Van C, Tsubokawa M, Bourne HR, Ramachandran J (1984) Amino acid sequence of retinal transducin at the site ADP- ribosylated by cholera toxin. J Biol Chem 259:696–698

Dudet LI, Chailler P, Dubreuil D, Martineau-Doize B (1996) *Pasteurella multocida* toxin stimulates mitogenesis and cytoskeleton reorganization in swiss 3T3 fibroblasts. J Cell Physiol 168:173–182

Essler M, Hermann K, Amano M, Kaibuchi K, Heesemann J, Weber PC, Aepfelbacher M (1998) *Pasteurella multocida* toxin increases endothelial permeability via rho kinase and myosin light chain phosphatase. J Immunol 161:5640–5646

Flatau G, Lemichez E, Gauthier M, Chardin P, Paris S, Fiorentini C, Boquet P (1997) Toxin-induced activation of the G protein p21 Rho by deamidation of glutamine. Nature 387:729–733

Frandsen PL, Foged NT, Petersen SK, Bording A (1991) Characterization of toxin from different strains of *Pasteurella multocida* serotype A and D. Zentralbl Veterinarmed B 38:345–352

Freissmuth M, Gilman AG (1989) Mutations of GS alpha designed to alter the reactivity of the protein with bacterial toxins. Substitutions at ARG187 result in loss of GTPase activity. J Biol Chem 264:21907–21914

Geissler B, Bonebrake A, Sheahan KL, Walker ME, Satchell KJ (2009) Genetic determination of essential residues of the *Vibrio cholerae* actin cross-linking domain reveals functional similarity with glutamine synthetases. Mol Microbiol 73:858–868

Geissler B, Tungekar R, Satchell KJ (2010) Identification of a conserved membrane localization domain within numerous large bacterial protein toxins. Proc Natl Acad Sci U S A 107: 5581–5586

Gierschik P (1992) ADP-ribosylation of signal-transducing guanine nucleotide-binding proteins by pertussis toxin. Curr Top Microbiol Immunol 175:69–98

Gilman AG (1987) G proteins: transducers of receptor-generated signals. Annu Rev Biochem 56:615–649

Hamm HE (1998) The many faces of G protein signaling. J Biol Chem 273:669–672

Hart MJ, Jiang X, Kozasa T, Roscoe W, Singer WD, Gilman AG, Sternweis PC, Bollag G (1998) Direct stimulation of the guanine nucleotide exchange activity of p115 RhoGEF by $G\alpha_{13}$. Science 280:2112–2114

Hawes BE, Touhara K, Kurose H, Lefkowitz RJ, Inglese J (1994) Determination of the G beta gamma-binding domain of phosducin. A regulatable modulator of G beta gamma signaling. J Biol Chem 269:29825–29830

Hepler JR, Gilman AG (1992) G proteins. Trends Biochem Sci 17:383–387

Higgins TE, Murphy AC, Staddon JM, Lax AJ, Rozengurt E (1992) *Pasteurella multocida* toxin is a potent inducer of anchorage-independent cell growth. Proc Natl Acad Sci U S A 89: 4240–4244

Hoffmann C, Schmidt G (2004) CNF and DNT. Rev Physiol Biochem Pharmacol 152:49–63

Hoffmann C, Pop M, Leemhuis J, Schirmer J, Aktories K, Schmidt G (2004) The Yersinia pseudotuberculosis cytotoxic necrotizing factor (CNFY) selectively activates RhoA. J Biol Chem 279:16026–16032

Hollinger S, Hepler JR (2002) Cellular regulation of RGS proteins: modulators and integrators of G protein signaling. Pharmacol Rev 54:527–559

Jetzt A, Howe JA, Horn MT, Maxwell E, Yin Z, Johnson D, Kumar CC (2003) Adenoviral-mediated expression of a kinase-dead mutant of Akt induces apoptosis selectively in tumor cells and suppresses tumor growth in mice. Cancer Res 63:6697–6706

Kalinec G, Nazarali AJ, Hermouet S, Xu N, Gutkind JS (1992) Mutated alpha subunit of the Gq protein induces malignant transformation in NIH 3T3 cells. Mol Cell Biol 12:4687–4693

Kamitani S, Kitadokoro K, Miyazawa M, Toshima H, Fukui A, Abe H, Miyake M, Horiguchi Y (2010) Characterization of the membrane-targeting C1 domain in *Pasteurella multocida* toxin. J Biol Chem 285:25467–25475

Kamitani S, Ao S, Toshima H, Tachibana T, Hashimoto M, Kitadokoro K, Fukui-Miyazaki A, Abe H, Horiguchi Y (2011) Enzymatic actions of *Pasteurella multocida* toxin detected by monoclonal antibodies recognizing the deamidated alpha subunit of the heterotrimeric GTPase G(q). FEBS J. 278:2702–2712

Katada T, Ui M (1982) Direct modification of the membrane adenylate cyclase system by islet-activating protein due to ADP-ribosylation of a membrane protein. Proc Natl Acad Sci U S A 79:3129–3133

Katada T, Oinuma M, Ui M (1986) Two guanine nucleotide-binding proteins in rat brain serving as the specific substrate of islet-activating protein, pertussis toxin. Interaction of the alpha-subunits with beta gamma-subunits in development of their biological activities. J Biol Chem 261:8182–8191

Kitadokoro K, Kamitani S, Miyazawa M, Hanajima-Ozawa M, Fukui A, Miyake M, Horiguchi Y (2007) Crystal structures reveal a thiol protease-like catalytic triad in the C-terminal region of *Pasteurella multocida* toxin. Proc Natl Acad Sci U S A 104:5139–5144

Kozasa T, Jiang X, Hart MJ, Sternweis PM, Singer WD, Gilman AG, Bollag G, Sternweis PC (1998) p115 RhoGEF, a GTPase activating protein for Galpha12 and Galpha13. Science 280:2109–2111

Lacerda HM, Lax AJ, Rozengurt E (1996) *Pasteurella multocida* toxin, a potent intracellularly acting mitogen, induces $p125^{FAK}$ and paxillin tyrosine phosphorylation, actin stress fiber formation, and focal contact assembly in Swiss 3T3 cells. J Biol Chem 271:439–445

Lee E, Taussig R, Gilman AG (1992) The G226A mutant of $G_{s\alpha}$ highlights the requirement for dissociation of G protein subunits. J Biol Chem 267:1212–1218

Lemichez E, Flatau G, Bruzzone M, Boquet P, Gauthier M (1997) Molecular localization of the *Escherichia coli* cytotoxic necrotizing factor CNF1 cell-binding and catalytic domains. Mol Microbiol 24:1061–1070

Lutz S, Freichel-Blomquist A, Yang Y, Rumenapp U, Jakobs KH, Schmidt M, Wieland T (2005) The guanine nucleotide exchange factor p63RhoGEF, a specific link between Gq/11-coupled receptor signaling and RhoA. J Biol Chem 280:11134–11139

Maier U, Babich A, Nurnberg B (1999) Roles of non-catalytic subunits in gbetagamma-induced activation of class I phosphoinositide 3-kinase isoforms beta and gamma. J Biol Chem 274: 29311–29317

Miyazawa M, Kitadokoro K, Kamitani S, Shime H, Horiguchi Y (2006) Crystallization and preliminary crystallographic studies of the *Pasteurella multocida* toxin catalytic domain. Acta Crystallograph Sect F Struct Biol Cryst Commun 62:906–908

Murayama T, Ui M (1983) Loss of the inhibitory function of the guanine nucleotide regulatory component of adenylate cyclase due to its ADP ribosylation by islet-activating protein, pertussis toxin, in adipocyte membranes. J Biol Chem 258(5):3319–3326

Northup JK, Sternweis PC, Smigel MDC, Ross EM, Gilman AG (1980) Purification of the regulatory component of adenylate cyclase. Proc Natl Acad Sci U S A 77:6516–6520

Nürnberg B (1997) Pertussis toxin as a cell biological tool. In: Aktories K (ed) Bacterial toxins—tools in cell biology and pharmacology. Chapman & Hall, Weinheim

Oldham WM, Hamm HE (2006) Structural basis of function in heterotrimeric G proteins. Q Rev Biophys 39:117–166

Oldham WM, Hamm HE (2008) Heterotrimeric G protein activation by G-protein-coupled receptors. Nat Rev Mol Cell Biol 9:60–71

Orth JH, Blöcker D, Aktories K (2003) His1205 and His 1223 are essential for the activity of the mitogenic *Pasteurella multocida* toxin. Biochemistry 42:4971–4977

Orth JH, Lang S, Aktories K (2004) Action of Pasteurella multocida toxin depends on the helical domain of Galphaq. J Biol Chem 279:34150–34155

Orth JH, Lang S, Taniguchi M, Aktories K (2005) Pasteurella multocida toxin-induced activation of RhoA is mediated via two families of G{alpha} proteins, G{alpha}q and G{alpha}12/13. J Biol Chem 280:36701–36707

Orth JH, Lang S, Preuss I, Milligan G, Aktories K (2007) Action of Pasteurella multocida toxin on Galpha(q) is persistent and independent of interaction with G-protein-coupled receptors. Cell Signal 19:2174–2182

Orth JH, Fester I, Preuss I, Agnoletto L, Wilson BA, Aktories K (2008) Activation of Galphai and subsequent uncoupling of receptor-Galphai signaling by Pasteurella multocida toxin. J Biol Chem 283:23288–23294

Orth JH, Preuss I, Fester I, Schlosser A, Wilson BA, Aktories K (2009) Pasteurella multocida toxin activation of heterotrimeric G proteins by deamidation. Proc Natl Acad Sci U S A 106:7179–7184

Parekh AB (2006) On the activation mechanism of store-operated calcium channels. Pflugers Arch 453:303–311

Peters PJ, Hsu VW, Ooi CE, Finazzi D, Teal SB, Oorschot V, Donaldson JG, Klausner RD (1995) Overexpression of wild-type and mutant ARF1 and ARF6: distinct perturbations of nonoverlapping membrane compartments. J Cell Biol 128:1003–1017

Petersen SK (1990) The complete nucleotide sequence of the *Pasteurella multocida* toxin gene and evidence for a transcriptional repressor. TxaR Mol Microbiol 4:821–830

Pettit RK, Ackermann MR, Rimler RB (1993) Receptor-mediated binding of *Pasteurella multocida* dermonecrotic toxin to canine osteosarcoma and monkey kidney (vero) cells. Labor Invest 69:94–100

Preuss I, Kurig B, Nürnberg B, Orth JH, Aktories K (2009) Pasteurella multocida toxin activates Gbetagamma dimers of heterotrimeric G proteins. Cell Signal 21:551–558

Pullinger GD, Lax AJ (2007) Histidine residues at the active site of the *Pasteurella multocida* toxin. Open Biochem J 1:7–11

Pullinger GD, Sowdhamini R, Lax AJ (2001) Localization of functional domains of the mitogenic toxin of *Pasteurella multocida*. Infect Immun 69:7839–7850

Radhika V, Dhanasekaran N (2001) Transforming G proteins. Oncogene 20:1607–1614

Rasmussen SG, DeVree BT, Zou Y, Kruse AC, Chung KY, Kobilka TS, Thian FS, Chae PS, Pardon E, Calinski D, Mathiesen JM, Shah ST, Lyons JA, Caffrey M, Gellman SH, Steyaert J, Skiniotis G, Weis WI, Sunahara RK, Kobilka BK (2011) Crystal structure of the beta2 adrenergic receptor-Gs protein complex. Nature 477:549–555

Repella TL, Ho M, Chong TPM, Bannai Y, Wilson BA (2011) Arf6-dependent intracellular trafficking of *Pasteurella multocida* toxin and pH-dependent translocation from late endosomes. Toxins 3:218–241

Riobo NA, Manning DR (2005) Receptors coupled to heterotrimeric G proteins of the G12 family. Trends Pharmacol Sci 26:146–154

Ross EM, Wilkie TM (2000) GTPase-activating proteins for heterotrimeric G proteins: regulators of G protein signaling (RGS) and RGS-like proteins. Annu Rev Biochem 69:795–827

Rozengurt E, Higgins T, Chanter N, Lax AJ, Staddon JM (1990) *Pasteurella multocida* toxin: potent mitogen for cultured fibroblasts. Proc Natl Acad Sci U S A 87:123–127

Schmidt G, Sehr P, Wilm M, Selzer J, Mann M, Aktories K (1997) Gln63 of Rho is deamidated by *Escherichia coli* cytotoxic necrotizing factor 1. Nature 387:725–729

Smyth MG, Pickersgill RW, Lax AJ (1995) The potent mitogen *Pasteurella multocida* toxin is highly resistant to proteolysis but becomes susceptible at lysosomal pH. FEBS Lett 360:62–66

Smyth MG, Sumner IG, Lax AJ (1999) Reduced pH causes structural changes in the potent mitogenic toxin of *Pasteurella multocida*. FEMS Microbiol Lett 180:15–20

Sprang SR (1997) G protein mechanisms: insights from structural analysis. Annu Rev Biochem 66:639–678

Staddon JM, Chanter N, Lax AJ, Higgins TE, Rozengurt E (1990) Pasteurella Multocida toxin, a potent mitogen, stimulates protein kinase C-dependent and -independent protein phosphorylation in swiss 3T3 cells. J Biol Chem 265:11841–11848

Staddon JM, Barker CJ, Murphy AC, Chanter N, Lax AJ, Michell RH, Rozengurt E (1991) *Pasteurella multocida* toxin, a potent mitogen, increases inositol 1,4,5-triphosphate and mobilizes Ca^{2+} in swiss 3T3 cells. J Biol Chem 266:4840–4847

Stephens LR, Eguinoa A, Erdjument-Bromage H, Lui M, Cooke F, Coadwell J, Smrcka AS, Thelen M, Cadwallader K, Tempst P, Hawkins PT (1997) The G beta gamma sensitivity of a PI3 K is dependent upon a tightly associated adaptor, p101. Cell 89:105–114

Stoll T, Markwirth G, Reipschlager S, Schmidt G (2009) A new member of a growing toxin family – *Escherichia coli* cytotoxic necrotizing factor 3 (CNF3). Toxicon 54:745–753

Takasaki J, Saito T, Taniguchi M, Kawasaki T, Moritani Y, Hayashi K, Kobori M (2004) A novel Galphaq/11-selective inhibitor. J Biol Chem 279:47438–47445

Taylor SJ, Smith JA, Exton JH (1990) Purification from bovine liver membranes of a guanine nucleotide-dependent activator of phosphoinositide-specific phospholipase C. Immunologic identification as a novel G-protein alpha subunit. J Biol Chem 265:17150–17156

Tesmer JJ, Berman DM, Gilman AG, Sprang SR (1997) Structure of RGS4 bound to AlF4-activated G(i alpha1): stabilization of the transition state for GTP hydrolysis. Cell 89:251–261

Ui M (1984) Islet-activating protein, pertussis toxin: a probe for functions of the inhibitory guanine nucleotide regulatory component of adenylate cyclase. Trendsss Pharmacol Sci 5:277–279

Van Raamsdonk CD, Bezrookove V, Green G, Bauer J, Gaugler L, O'Brien JM, Simpson EM, Barsh GS, Bastian BC (2008) Frequent somatic mutations of GNAQ in uveal melanoma and blue naevi. Nature 457:599–602

Van Sandvig K, Deurs B (2005) Delivery into cells: lessons learned from plant and bacterial toxins. Gene Ther 12:865–872

Vogt S, Grosse R, Schultz G, Offermanns S (2003) Receptor-dependent RhoA activation in G12/G13-deficient cells. J Biol Chem 278:28743–28749

Ward PN, Miles AJ, Sumner IG, Thomas LH, Lax AJ (1998) Activity of the mitogenic *Pasteurella multocida* toxin requires an essential C-terminal residue. Infect Immun 66: 5636–5642

Westfield GH, Rasmussen SG, Su M, Dutta S, DeVree BT, Chung KY, Calinski D, Velez-Ruiz G, Oleskie AN, Pardon E, Chae PS, Liu T, Li S, Woods VL Jr, Steyaert J, Kobilka BK, Sunahara

RK, Skiniotis G (2011) Structural flexibility of the G alpha s alpha-helical domain in the beta2-adrenoceptor Gs complex. Proc Natl Acad Sci U S A 108:16086–16091

Williamson MP (1994) The structure and function of proline-rich regions in proteins. Biochem J 297(Pt 2):249–260

Williamson EA, Ince PG, Harrison D, Kendall-Taylor P, Harris PE (1995) G-protein mutations in human pituitary adrenocorticotrophic hormone-secreting adenomas. Eur J Clin Invest 25:128–131

Wilson BA, Zhu X, Ho M, Lu L (1997) *Pasteurella multocida* toxin activates the inositol triphosphate signaling pathway in Xenopus oocytes via $G_{q\alpha}$-coupled phospholipase C-β1. J Biol Chem 272:1268–1275

Wu G, Benovic JL, Hildebrandt JD, Lanier SM (1998) Receptor docking sites for G-protein betagamma subunits. implications for signal regulation. J Biol Chem 273:7197–7200

Zywietz A, Gohla A, Schmelz M, Schultz G, Offermanns S (2001) Pleiotropic effects of *Pasteurella multocida* toxin are mediated by Gq-dependent and -independent mechanisms. Involvement of Gq but not G11. J Biol Chem 276:3840–3845

Pasteurella multocida Toxin Interaction with Host Cells: Entry and Cellular Effects

Brenda A. Wilson and Mengfei Ho

Abstract The mitogenic dermonecrotic toxin from *Pasteurella multocida* (PMT) is a 1285-residue multipartite protein that belongs to the A-B family of bacterial protein toxins. Through its G-protein-deamidating activity on the α subunits of heterotrimeric G_q-, G_i- and $G_{12/13}$-proteins, PMT potently stimulates downstream mitogenic, calcium, and cytoskeletal signaling pathways. These activities lead to pleiotropic effects in different cell types, which ultimately result in cellular proliferation, while inhibiting cellular differentiation, and account for the myriad of physiological outcomes observed during infection with toxinogenic strains of *P. multocida*.

Contents

B. A. Wilson (✉) · M. Ho
Department of Microbiology and the Host-Microbe Systems Theme of the Institute for Genomic Biology, University of Illinois at Urbana-Champaign, Urbana, IL 61801, USA
e-mail: bawilson@life.illinois.edu
URL: http://mcb.illinois.edu/faculty/profile/wilson7

Current Topics in Microbiology and Immunology (2012) 361: 93–111
DOI: 10.1007/82_2012_219

Published Online: 4 May 2012

1 Introduction

Toxinogenic strains of *Pasteurella multocida*, mostly serogroup D and some A, are associated with a number of epizootic and zoonotic diseases, including dermonecrosis, pasteurellosis, and atrophic rhinitis, in domestic and wild animals (DiGiacomo et al. 1991b; Foged 1992; Frymus et al. 1991; Kielstein 1986; Wilson and Ho 2006) and in humans who come in close contact with infected animals (Arashima and Kumasaka 2005; Donnio et al. 1991; Donnio et al. 2004; Garcia 1997; Iaria and Cascio 2007; Kobayaa et al. 2009; Wilson and Ho 2006). The major virulence factor responsible for the symptoms manifested during infection with these *P. multocida* strains is a 1285-residue (146-kDa) protein toxin (PMT), which belongs to the large, prominent group of intracellularly acting, multipartite A-B toxins that modify eukaryotic G-proteins (Wilson and Ho 2010). A-B toxins bind to host cell receptors through their binding B domains and facilitate the cellular uptake and delivery (translocation) of their toxic activity A domains into the host cell cytosol, where the A domains then interact with and modify their cellular G-protein targets to cause cellular toxicity.

The G-protein targets of A-B toxins are GTPases that regulate various cellular signal transduction pathways by cycling between an inactive GDP-bound form and an active GTP-bound form. PMT selectively deamidates a key active site Gln residue of the α subunit of its heterotrimeric G-protein targets, G_q, G_i , and $G_{12/13}$ (Orth et al. 2009). This modification locks the GTPase activity of the α subunit into an active state, resulting in persistent stimulation of downstream signaling pathways modulated by the G-protein targets [reviewed in (Wilson and Ho 2010, 2011)]. While we are beginning to have a clearer picture of the molecular basis for the biochemical activity of PMT, much less is known about the molecular mechanisms of cellular intoxication or how the selective deamidation of its G-protein targets leads to the myriad of cellular outcomes observed. In this review, we focus on our current understanding of how PMT interacts with host cells to gain entry and elicit various cellular effects through its G-protein deamidase activity.

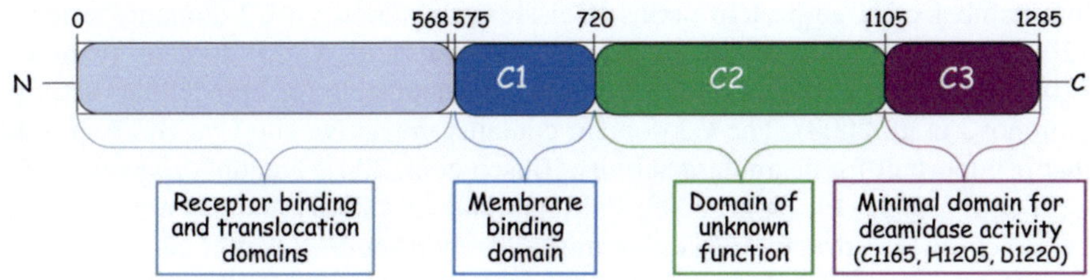

Fig. 1 A schematic diagram of the overall structure of PMT. PMT-N (residues 1–568) and the known C-terminal structural domains (C1–C3) of PMT-C (residues 569–1285) are indicated along with their known or putative functions

2 PMT Structure and Function

PMT is a member of the dermonecrotic toxin family, which includes the cytotoxic necrotizing factors from *E. coli* (CNF1, CNF2, and CNF3) and *Yersinia pseudotuberculosis* (CNFy) and the dermonecrotic toxin from *Bordetella* species (DNT) (Aktories and Barbieri 2005; Hoffmann and Schmidt 2004; Wilson and Ho 2010). Members of this family of A-B toxins share with each other sequence and structural features that enable them to enter host cells and then gain access to their G-protein targets and modify them. The N-terminus of PMT (PMT-N) has significant sequence similarity with the N-termini of the CNFs (Buys et al. 1990; Falbo et al. 1993; Kamps et al. 1990; Lockman et al. 2002; Oswald et al. 1994; Petersen and Foged 1989; Stoll et al. 2009) and to a lesser extent that of DNT (Pullinger et al. 1996). Although there is no crystal structure available for any of the full-length dermonecrotic toxins such that the actual domains responsible for receptor binding and translocation have not yet been clearly defined, there is some biochemical evidence that the N-termini of these proteins are indeed important for toxin binding and translocation (Baldwin et al. 2004; Blumenthal et al. 2007; Brothers et al. 2011; Chung et al. 2003; Kim et al. 2005; Lemichez et al. 1997; Pullinger et al. 2001).

The CNFs and DNT share over 50% sequence similarity in their C-terminal domains (residues 720–1014 in the CNFs, 1176–1464 in DNT), which have deamidase and/or transglutaminase activity (Hoffmann and Schmidt 2004). Their common G-protein targets belong to the Rho family of small GTPases, such as RhoA, Rac1, and Cdc42, involved in regulation of cytoskeletal function (Aktories and Barbieri 2005). The G-protein deamidase activity of PMT responsible for activation of mitogenic and calcium signaling pathways also resides within the C-terminal 700 amino acids of PMT (PMT-C) (Baldwin et al. 2004; Busch et al. 2001; Orth et al. 2003; Orth et al. 2009; Pullinger and Lax 2007; Pullinger et al. 2001). The crystal structures of PMT-C [PDB 2EBF] (Kitadokoro et al. 2007) and the C-terminal deamidase domain (residues 720–1014) of CNF1 [PDB 1HQ0] (Buetow et al. 2001) are available. The crystal structure of PMT-C (Kitadokoro et al. 2007) revealed three distinct domains (Fig. 1): a C1 domain (residues 575–719) that has sequence and structural homology with the membrane-targeting domains found in a number of large protein toxins (Geissler et al. 2010); a C2 domain (residues 720–1104) that is as-of-yet unknown function; and a C3 domain (residues 1105–1285) that harbors the minimal domain responsible for intracellular activity (Aminova et al. 2008). The C3 domain contains the active site Cys-His-Asp triad that is important for deamidase activity (Busch et al. 2001; Kitadokoro et al. 2007; Orth et al. 2003; Orth et al. 2009; Pullinger and Lax 2007) and has a papain-like cysteine protease structural fold that most closely resembles that of certain protein transglutaminases (Kitadokoro et al. 2007; Wilson and Ho 2010). Of particular note, however, was the surprising finding that PMT and CNF1 catalyze the same enzymatic reaction on a functionally equivalent Gln residue at the active site of their respective substrates (Gln-61 in Rac1 and Cdc42; Gln-63 in RhoA; Gln-205 in $G\alpha_i$; Gln-209 in $G\alpha_q$; Gln-229 in $G\alpha_{12/13}$) (Flatau et al. 1997; Orth et al. 2009; Schmidt et al. 1997) and have essential active site catalytic His and Cys residues (Buetow et al. 2001; Busch et al. 2001; Kitadokoro et al. 2007; Orth et al. 2003; Schmidt et al. 1998). Yet, there is no discernable sequence or structural similarity between PMT-C3 and the catalytic domain of CNF1 (Wilson and Ho 2010).

3 Cellular Uptake of PMT

Little is known about the cellular intoxication mechanisms of PMT. As is the case for the other dermonecrotic toxin family members (Hoffmann and Schmidt 2004), PMT-N contains the functional receptor-binding and translocation domains. Although the precise boundaries for the binding and translocation domains have not yet been defined, PMT-N appears to be sufficient for mediating cell binding, uptake, and subsequent delivery of PMT-C into the cytosol (Baldwin et al. 2004; Brothers et al. 2011; Pullinger et al. 2001). Indeed, a fusion protein of PMT-N with GFP at the C-terminus also binds and enters cells (Repella et al. 2011). We will now consider what is known about each of these steps in PMT intoxication.

3.1 Binding to Host Cells

A-B toxins are known to employ different types of receptors to intoxicate host cells. For example, cholera and pertussis toxins utilize gangliosides alone (Merritt et al. 1994; Stein et al. 1994), diphtheria toxin uses proteins alone (Naglich et al. 1992), and botulinum and tetanus neurotoxins use both proteins and gangliosides as co-receptors (Binz and Rummel 2009; Dong et al. 2006). Earlier reports suggested that PMT utilized gangliosides, namely GM_1, GM_2 and GM_3, as receptors (Dudet et al. 1996; Pettit et al. 1993a). However, a more recent study using several different methods, including PMT binding to membrane lipid components using TLC-overlay, liposome-pulldown, and surface plasmon resonance (SPR) experiments, showed that PMT and PMT-N do not bind gangliosides (GM_1, GM_2 or GM_3), but instead both bound well to asialogangliosides, such as lactosylceramide, and in particular sphingomyelin (SM) and positively charged membrane phospholipids, such as phosphatidylcholine (PC) and to some extent phosphatidylethanolamine (PE) and some other lipid components (Brothers et al. 2011).

Results from this study also implicated the potential involvement of a protein co-receptor in PMT binding to cells (Brothers et al. 2011), since no single treatment with sphingomyelinase, phospholipase D, or trypsin completely abolished PMT binding to cell membranes, and instead each treatment reduced cellular binding of PMT by about one-third. Indeed, trypsin treatment of cell membranes had minimal effect on membranes already depleted of choline or phosphocholine head groups, indicating that protein binding is only important for PMT binding when PMT is also interacting with SM and/or PC in the membrane. Based on the observed binding kinetics from the SPR studies under the different treatment conditions, a model for PMT interaction with host cells was proposed, whereby PMT initially binds nonspecifically and with low affinity to the more abundant membrane lipid components such as PC at the surface, but this is then followed by a more specific, tight-binding interaction with SM and possibly other membrane components, including a putative protein co-receptor(s).

3.2 Trafficking in Cells

Once PMT binds to host cells, it is internalized through receptor-mediated endocytosis and then trafficked to acidic endosomes, where it is translocated across the vesicle membrane into the host cell cytosol (Baldwin et al. 2004; Repella et al. 2011; Rozengurt et al. 1990). Although knowledge of the detailed mechanism of this process is lacking, we are beginning to gain some insights. A recent study showed that receptor-mediated endocytosis and initial trafficking of PMT are dependent on the small regulatory G-protein Arf6 (Repella et al. 2011). PMT is initially internalized and trafficked to Arf6-containing vesicles, where it co-localizes with cholera toxin and transferrin, but transferrin is subsequently trafficked to recycling endosomes and cholera toxin is trafficked retrograde to the endoplasmic reticulum, while PMT-containing early endosomes are diverted to late endosomes.

Disassembly of microtubules important for trafficking from early to late endosomes by treatment with nocodazole or disruption of actin polymerization by treatment with cytochalasin D also block PMT activation of mitogenic signaling (Repella et al. 2011), suggesting that membrane translocation and cytotoxicity of PMT are dependent on trafficking to late acidic endosomes. Additional evidence that PMT trafficking to late acidic endosomes is important for PMT action comes from the finding that treatment with brefeldin A, which disrupts Golgi-endoplasmic reticulum trafficking, enhances PMT activity by over 20-fold, presumably by preventing PMT trafficking through nonproductive pathways that do not lead to translocation (Repella et al. 2011). Based on these results, a model was proposed for PMT intoxication, as depicted in Fig. 2.

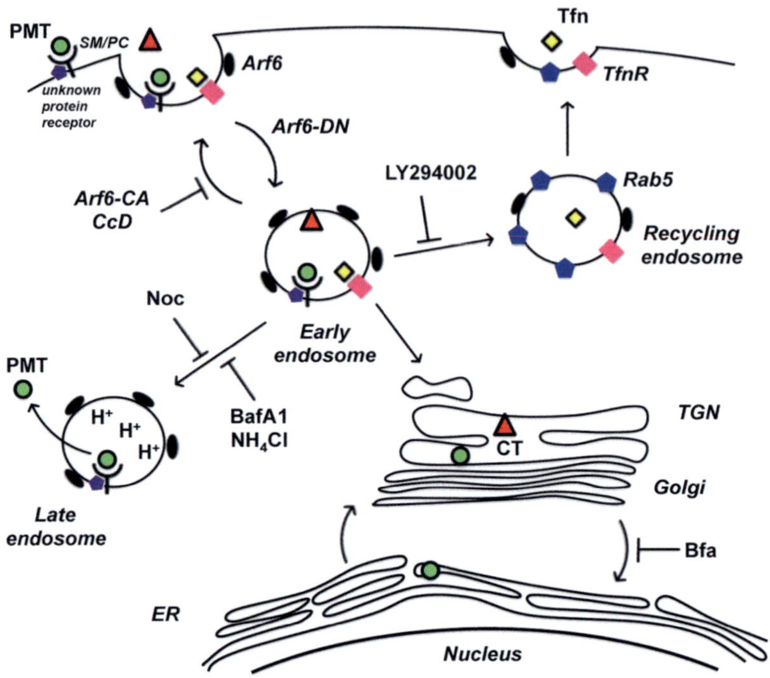

Fig. 2 Proposed model of PMT entry and trafficking. *Tfn* transferrin, *TfnR* transferrin receptor, *CT* cholera toxin, *Rab5* GTPase marker of recycling endosomes, *TGN* trans-Golgi network, *ER* endoplasmic reticulum, *SM/PC* sphingomyelin/PC receptors of PMT, *Arf6-CA* constitutively active Arf6 GTPase, *Arf6-DN* dominantly negative Arf6, *CcD* cytochalasin D, *LY294002* PI3 K inhibitor of early endosome-recycling endosome fusion, *Noc* nocodazole, *BafA1* bafilomycin A1, *Bfa* brefeldin A. [Adapted from (Repella et al. 2011)]

3.3 Translocation Across Cell Membranes

The above results showing that PMT activity depends on trafficking to late acidic endosomes, where presumably translocation into the cytosol occurs, confirmed earlier findings that weak bases, such as ammonium chloride, chloroquine, or methylamine, which buffer acidification of endosomes, inhibit PMT effects on cells (Rozengurt et al. 1990). Moreover, bafilomycin A1, a potent and specific inhibitor of the vacuolar H^+-ATPase pump that is responsible for acidifying early to late endosomes, likewise inhibits PMT action (Baldwin et al. 2004; Repella et al. 2011). Further evidence for a low pH-dependent membrane translocation event in PMT action was provided by experiments that showed cell surface-bound PMT could directly enter cells, even in the presence of bafilomycin A1, through a low pH pulse at 4°C, which normally blocks endocytosis (Baldwin et al. 2004).

A predicted helix-loop-helix motif, which corresponded to a similar helix-loop-helix in CNF1 (Pei et al. 2001), was identified in PMT (Baldwin et al. 2004).

This motif contained two hydrophobic helices (residues 402–423 and 437–457) linked by a hydrophilic loop (residues 424–436) and was proposed to be involved in the pH-sensitive membrane translocation step (Baldwin et al. 2004). Mutation of acidic residues (Asp-373 and Asp-379) in the loop region of the helix-loop-helix motif in CNF1 resulted in complete loss of biological activity (Pei et al. 2001). However, mutation of analogous residues in the loop region of PMT (Asp-425, Asp-432, Glu-434) resulted in only partial reduction of toxin activity, whereas mutation of an acidic residue (Asp-401) just outside of the predicted motif completely abolished PMT activity (Baldwin et al. 2004). Nevertheless, these results support a model where this helix-loop-helix region of PMT is part of a putative translocation domain.

4 Effects on Cell Signaling

PMT causes a number of pleiotropic effects on targeted host cells. Most notably, PMT induces strong mitogenic (Aminova et al. 2008; Dudet et al. 1996; Mullan and Lax 1996, 1998; Rozengurt et al. 1990; Seo et al. 2000; Wilson et al. 2000; Zywietz et al. 2001) and anti-apoptotic (Aminova and Wilson 2007; Orth et al. 2007a; Preuss et al. 2010; Sabri et al. 2002) signaling in various cell lines, while simultaneously downregulating signaling pathways involved in cellular differentiation, including osteogenesis (Harmey et al. 2004; Mullan and Lax 1998; Sterner-Kock et al. 1995), adipogenesis (Aminova and Wilson 2007), and immune cell differentiation (Bagley et al. 2005; Blocker et al. 2006; Jordan et al. 2003; van Diemen et al. 1994, 1996). The cellular outcomes of these effects of PMT are manifested in the various observed disease symptoms at the different sites of infection.

Exposure to PMT through respiratory infection results in bone resorption of nasal turbinates in progressive atrophic rhinitis (DiGiacomo et al. 1991b; Foged 1992; Frymus et al. 1991; Lax and Chanter 1990; Magyar 1989; Wilson and Ho 2006) and pneumonia (a.k.a, pasteurellosis in rabbits or bovine respiratory distress in cattle) (Chrisp and Foged 1991; DiGiacomo et al. 1991a; Frymus et al. 1991; Kielstein 1986; Klein and Cunha 1997). Chronic or systemic infections also lead to testicular and splenic atrophy (Chrisp and Foged 1991; Nakai et al. 1984), liver necrosis (Cheville and Rimler 1989; Cheville et al. 1988; Chrisp and Foged 1991), kidney or bladder impairment (Hoskins et al. 1997), overall reduced weight and body fat (Cheville and Rimler 1989; Thurston et al. 1992), and growth retardation (Ackermann et al. 1996; Ackermann et al. 1995; Al-Haddawi et al. 2001). Infections in humans usually result from close contact with infected animals, particularly through respiratory exposure or bite wounds and have similar disease manifestations as those observed in animals (Arashima and Kumasaka 2005; Donnio et al. 1999; Donnio et al. 1991; Donnio et al. 2004; Frederiksen 1993; Garcia 1997; Griego et al. 1995; Henderson et al. 2010; Holst et al. 1992; Iaria and

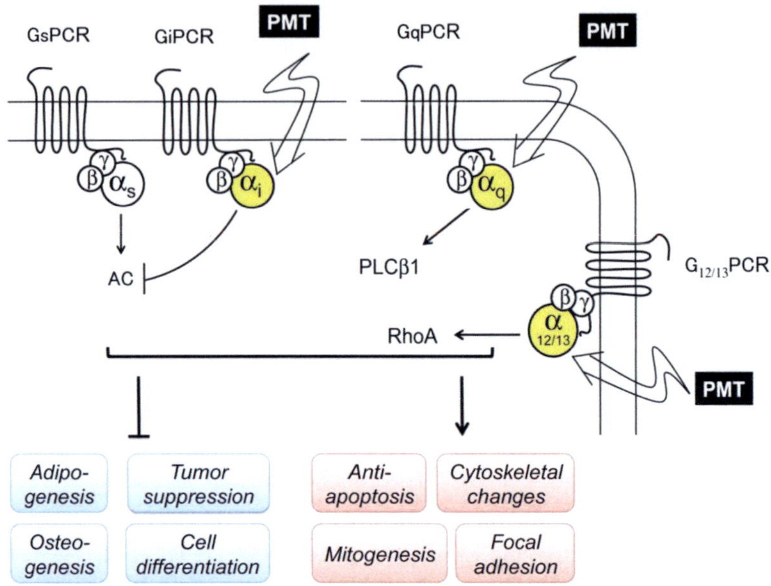

Fig. 3 Signaling pathways modulated by PMT and their cellular outcomes

Cascio 2007; Kobayaa et al. 2009; Migliore et al. 2009; Satomura et al. 2010; Waldor et al. 1992).

4.1 Cellular Responses

Cellular responses to PMT are induced by activation of at least three different heterotrimeric G-protein families, G_q, G_i, and $G_{12/13}$ (Fig. 3). PMT-mediated activation of G_q and $G_{12/13}$ signaling leads to stimulation of mitogenic responses through increased intracellular calcium and inositol phosphate levels as a result of activation of phospholipase Cβ (PLCβ) by G_q (Aminova et al. 2008; Aminova and Wilson 2007; Luo et al. 2008; Murphy and Rozengurt 1992; Staddon et al. 1991; Wilson et al. 1997) and cytoskeletal changes through activation of Rho-dependent actin signaling by G_q and $G_{12/13}$ (Aepfelbacher and Essler 2001; Lacerda et al. 1996; Orth et al. 2005; Sagi et al. 2001). Simultaneous activation of G_i signaling by PMT leads to blockade of G_s-regulated adenylyl cyclase (AC) activity (Orth et al. 2008), which in turn downregulates cAMP-dependent signaling pathways involved in cellular differentiation processes. In the following sections we will summarize the known cellular effects mediated through signaling pathways activated by PMT action on its G-protein targets. It should also be noted that in addition to release of the activated Gα subunits, PMT action also releases the cognate G$\beta\gamma$ subunits from the heterotrimeric complexes, which concomitantly modulate other signaling pathways such as the phosphoinositide 3-kinase γ (PI3 Kγ) pathway (Preuss et al. 2009) and perhaps ion channels (Bunemann et al. 2000; Meyer et al. 2001).

4.2 Calcium Signaling

PMT strongly stimulates phospholipase Cβ1 and to a lesser extent PLCβ3, but not PLCβ2 (Wilson et al. 1997), through its selective activation of Gα_q protein and not the closely related Gα_{11} protein (Staddon et al. 1991; Wilson et al. 1997; Zywietz et al. 2001). Although the mechanism whereby PMT discriminates between Gα_q and Gα_{11} is not clear, this preference appears to occur through selective recognition of the helical domain of the α subunits, and not the highly conserved regions flanking the target Gln-209 residue (Orth et al. 2004). Dissociation of the Gα_q subunit from the G$\alpha\beta\gamma$ complex, through treatment with anti-Gβ antibodies to release the Gα_q subunit or by treatment with pertussis toxin to sequester the G$\beta\gamma$ subunits away from the Gα_q subunit, potentiates the PMT-induced PLCβ response (Wilson et al. 1997). Overexpression of Gα_q protein likewise enhances the PMT-induced response. These results support the monomeric form of Gα_q protein as the preferred substrate of PMT. PMT deamidation of Gα_q results in constitutive activation (Orth et al. 2009), which is irreversible and persistent (Orth et al. 2007b; Wilson et al. 1997) and independent of interaction with G-protein-coupled receptors (GPCRs) (Orth et al. 2008; Orth et al. 2007b; Orth et al. 2009; Wilson et al. 1997).

Activation of PLCβ1 triggers the hydrolysis of phosphatidylinositol 4,5-bisphosphate to release inositol 1,3,5-trisphosphate (IP_3) and diacylglycerol, which in turn results in mobilization of intracellular calcium pools (Staddon et al. 1991; Wilson et al. 1997) and stimulates calcium signaling pathways (Aminova et al. 2008; Aminova and Wilson 2007; Hennig et al. 2008; Luo et al. 2008) and protein kinase C (PCK)-dependent and -independent phosphorylations (Obreztchikova et al. 2006; Orth et al. 2007a; Ozgen et al. 2008; Sabri et al. 2002; Seo et al. 2000; Staddon et al. 1990; Thomas et al. 2001; Wilson et al. 2000; Zywietz et al. 2001).

4.3 Mitogenic Signaling

PMT elicits a strong mitogenic response in most cultured cell lines that leads to either proliferative or cytopathic effects. In fibroblasts (Dudet et al. 1996; Rozengurt et al. 1990; Sabri et al. 2002; Seo et al. 2000; Wilson et al. 2000; Zywietz et al. 2001), preadipocytes (Aminova and Wilson 2007), and osteoclasts (Felix et al. 1992; Gwaltney et al. 1997; Hildebrand et al. 2010a; Jutras and Martineau-Doize 1996; Martineau-Doize et al. 1993; Mullan and Lax 1996, 1998),

PMT potently stimulates DNA synthesis and proliferation. PMT also strongly stimulates anti-apoptotic signaling pathways (Hildebrand et al. 2010b; Orth et al. 2007a; Ozgen et al. 2008; Preuss et al. 2010; Staddon et al. 1992). In other cells, such as embryonic bovine lung cells (Rutter and Luther 1984), Vero cells (Pennings and Storm 1984; Wilson et al. 2000), cardiomyocytes (Obreztchikova et al. 2006; Sabri et al. 2000; Sabri et al. 2002), osteoblasts, and osteosarcoma cells (Gwaltney et al. 1997; Harmey et al. 2004; Pettit et al. 1993a; Sterner-Kock et al. 1995), PMT elicits a cytopathic response characterized by stress responses, actin rearrangements and other morphological changes (see below). However, after the strong initial PMT-induced cellular response, no further stimulation occurs upon additional treatment with PMT, indicating that an uncoupling of the G-protein signaling pathways occurs (Orth et al. 2008; Wilson et al. 2000; Wilson et al. 1997).

4.4 Cytoskeletal Signaling

PMT initiates RhoA-dependent cytoskeletal signaling, including actin rearrangements, stress fiber formation, and focal adhesion assembly (Aepfelbacher and Essler 2001; Blocker et al. 2006; Dudet et al. 1996; Lacerda et al. 1996; Ohnishi et al. 1998; Orth et al. 2005; Sabri et al. 2002; Sagi et al. 2001; Thomas et al. 2001). However, PMT does not directly modify RhoA (Horiguchi 2001; Lacerda et al. 1996; Ohnishi et al. 1998); instead, PMT activation of RhoA occurs indirectly through activation of $G\alpha_{12/13}$ (Orth et al. 2005) and to some extent $G\alpha_q$ (Sagi et al. 2001). PMT-induced RhoA activation leads to activation of Rho kinase and phosphorylation of focal adhesion kinase (Lacerda et al. 1996; Thomas et al. 2001) and myosin light chain (Aepfelbacher and Essler 2001; Essler et al. 1998), which in turn regulates the actin cytoskeleton, stress fiber formation, focal adhesion assembly, and endothelial cell barrier permeability. PMT disturbance of endothelial barrier function has been attributed as the cause of the observed vascular effects in dermonecrotic lesions from PMT-infected bite wounds (Aepfelbacher and Essler 2001; Elling et al. 1988).

4.5 cAMP Signaling

While potently inducing mitogenic and cytoskeletal signaling through activation of $G\alpha_q$ and $G\alpha_{12/13}$ signaling, PMT simultaneously inhibits AC activity and downstream cAMP-mediated processes through activation of $G\alpha_i$ signaling (Orth et al. 2008). PMT-induced $G\alpha_i$ activation locks the $G\alpha_i$ subunit in its monomeric active form and overrides isoproterenol stimulation of AC by G_s-protein-coupled receptors (Orth et al. 2008). In addition, PMT deamidation of $G\alpha_i$ interferes with the interaction of the $G\alpha_i$ subunit and its cognate $G\beta\gamma$ subunits, and thereby converts the $G\alpha_i$ protein into a pertussis toxin-insensitive form (Orth et al. 2009), since pertussis toxin prefers to ADP-ribosylate the heterotrimeric $G\alpha_i\beta\gamma$ complex and not the monomeric $G\alpha_i$ protein (Katada et al. 1986). Thus, PMT treatment effectively shifts the equilibrium to dissociate the heterotrimeric complex and releases $G\alpha_i$, which blocks AC and cAMP accumulation (Orth et al. 2008), and $G\beta\gamma$ subunits, which can interact with their cognate downstream effectors, such as PI3 Kγ (Preuss et al. 2009), which leads to inhibition of apoptotic signaling pathways (Preuss et al. 2010).

4.6 Adipogenic Signaling

PMT treatment was shown to prevent adipocyte differentiation and block adipogenesis in cultured 3T3-L1 cells under differentiation-inducing conditions (Aminova and Wilson 2007). In this study, PMT prevented expression of key adipocyte-specific markers, C/EBPα and PPARγ, in 3T3-L1 preadipocytes and downregulated these markers in mature adipocytes. PMT also prevented the downregulation of Pref1 (also called Dlk1), an EGF-like transmembrane protein that is strongly downregulated during adipocyte differentiation (Boney et al. 1996; Garces et al. 1999; Sul 2009). PMT was further shown to completely downregulate Notch1 mRNA and protein expression, while stabilizing β-catenin protein levels (Aminova and Wilson 2007). Notch1 and Wnt/β-catenin signaling pathways are involved in pivotal cell fate decisions (Andersson et al. 2011). Interestingly, the inhibitory effects of PMT on adipocyte differentiation and Notch1 could not be reversed by treatment with cyclosporine A (CsA) (Aminova and Wilson 2007), an inhibitor of calcium-calmodulin-dependent calcineurin signaling that is known to reverse G_q-PLCβ1-mediated inhibition of adipogenesis (Liu and Clipstone 2007; Neal and Clipstone 2002). These results suggest that PMT-induced G_q–PLCβ1 activation of calcium signaling is not the only signaling pathway mediated by PMT to block adipocyte differentiation, which leaves the possibility that PMT blockade of adipogenesis might also be mediated through PMT activation of either G_i or $G_{12/13}$ signaling.

4.7 Osteogenic Signaling

PMT is the primary etiological agent responsible for progressive atrophic rhinitis in pigs, rabbits, and other animals (Deeb et al. 1990; DiGiacomo et al. 1993; Foged 1992; Magyar 1989; Wilson and Ho 2006). Atrophic rhinitis is characterized by destruction of the nasal turbinate bones through disruption of bone biogenesis by osteoblasts and bone degradation processes (resorption) by macrophage-like osteoclasts (Kimman and Kamp 1986; Kimman et al. 1987; Mullan and Lax 1996, 1998). In vivo, PMT treatment appears to promote bone resorption through differentiation of preosteoclasts into osteoclasts, which then proliferate (Jutras and Martineau-Doize 1996; Martineau-Doize et al. 1993). In cell culture, PMT also promotes bone resorption by osteoclasts (Felix et al. 1992; Gwaltney et al. 1997; Mullan and Lax 1996) and inhibits bone biogenesis by inhibiting osteoblast differentiation (Gwaltney et al. 1997; Harmey et al. 2004; Mullan and Lax 1998; Sterner-Kock et al. 1995).

4.8 Immune Signaling

Although immunization with PMT protein toxoid affords protection against atrophic rhinitis (Bourdon et al. 2007; Chanter and Rutter 1990; Foged et al. 1989; Frymus et al. 1989; Pettit et al. 1993b; Suckow 2000; Suckow et al. 1995; Thurston et al. 1992), natural infection with toxinogenic *P. multocida* is characterized by an overall lack of immune response against PMT (Bagley et al. 2005; Hamilton et al. 1998; Jordan et al. 2003; van Diemen et al. 1996). Indeed, in vivo PMT is a poor immunogen and appears to suppress antibody responses to PMT and other antigens (Bagley et al. 2005; Hamilton et al. 1998; Jordan et al. 2003), suggesting a possible in vivo role for PMT as an immunomodulator during infection.

While PMT activates human monocyte-derived and murine bone marrow-derived dendritic cells in vitro, it inhibits migration of the dendritic cells (Bagley et al. 2005; Blocker et al. 2006). PMT treatment results in G_q-dependent phosphorylation and activation of Janus tyrosine kinases (JAK1 and JAK2), which leads to activation of downstream STAT signaling and consequent upregulation of proinflammatory responses (Orth et al. 2007a) and cytokine signaling (Hildebrand et al. 2010b). Clearly, additional studies are needed to clarify the contrasting effects observed regarding PMT action on immune signaling in vitro versus in vivo.

5 Perspective

Initial studies of how bacterial protein toxins modulate host cells focused primarily on the structural organization and mode of action of the toxins. Recently, the focus of toxin studies has shifted more toward understanding the molecular interactions of the toxins with host cells, specifically how they are taken up, trafficked, and translocated into the cytosol, and how they modulate various cellular signaling pathways that lead to changes in cellular function and physiology. There are still a number of large gaps in our understanding of the intoxication process, not only for PMT but also for many other large AB toxins as well.

Since the active site Gln residue of the Gα-protein targeted by PMT is highly conserved throughout the heterotrimeric Gα-proteins, it remains to be determined which of the other Gα-proteins are also substrates for PMT and what the substrate recognition determinants are that discriminate one substrate from the others. Another area that remains unclear is the consequences to the Gα-protein once it has been modified by PMT, particularly the mechanism of its subsequent downregulation. Defining more clearly the substrate preferences of PMT and the ultimate consequences that toxin modification have on the various Gα-protein targets will assist in deciphering the differential effects on the signaling pathways elicited in different cell types and the overall cellular outcomes. Understanding toxin-induced changes in modulation of various cellular processes, including processes that facilitate actin rearrangements, proliferation, and cellular differentiation, will illuminate the critical role of toxins for the successful survival of the pathogen in the animal host. Certainly, reconciling in vitro observations with in vivo outcomes of toxin exposure is high on the list of tasks for future studies.

Acknowledgments Some of the work reported here was supported by grants from the National Institutes of Health (NIH/NIAID AI038396) and the US Department of Agriculture (NRI 1999-02295) (to B.A.W.).

References

Ackermann MR, Register KB, Stabel JR, Gwaltney SM, Howe TS, Rimler RB (1996) Effect of *Pasteurella multocida* toxin on physeal growth in young pigs. Am J Vet Res 57:848–852

Ackermann MR, Stabel JR, Pettit RK, Jacobson CD, Elmquist JK, Register KB, Rimler RB, Hilton JH (1995) Reduced physeal area and chondrocyte proliferation in *Pasteurella multocida* toxin-treated rats. Vet Pathol 32:674–682

Aepfelbacher M, Essler M (2001) Disturbance of endothelial barrier function by bacterial toxins and atherogenic mediators: a role for Rho/Rho kinase. Cell Microbiol 3:649–658

Aktories K, Barbieri JT (2005) Bacterial cytotoxins: targeting eukaryotic switches. Nat Rev Microbiol 3:397–410

Al-Haddawi MH, Jasni S, Israf DA, Zamri-Saad M, Mutalib AR, Sheikh-Omar AR (2001) Ultrastructural pathology of nasal and tracheal mucosa of rabbits experimentally infected with *Pasteurella multocida* serotype D:1. Res Vet Sci 70:191–197

Aminova LR, Luo S, Bannai Y, Ho M, Wilson BA (2008) The C3 domain of *Pasteurella multocida* toxin is the minimal domain responsible for activation of Gq-dependent calcium and mitogenic signaling. Protein Sci 17:945–949

Aminova LR, Wilson BA (2007) Calcineurin-independent inhibition of 3T3-L1 adipogenesis by *Pasteurella multocida* toxin: suppression of Notch1, stabilization of beta-catenin and pre-adipocyte factor 1. Cell Microbiol 9:2485–2496

Andersson ER, Sandberg R, Lendahl U (2011) Notch signaling: simplicity in design, versatility in function. Development 138:3593–3612

Arashima Y, Kumasaka K (2005) Pasteurellosis as zoonosis. Intern Med 44:692–693

Bagley KC, Abdelwahab SF, Tuskan RG, Lewis GK (2005) *Pasteurella multocida* toxin activates human monocyte-derived and murine bone marrow-derived dendritic cells in vitro but suppresses antibody production in vivo. Infect Immun 73:413–421

Baldwin MR, Lakey JH, Lax AJ (2004) Identification and characterization of the *Pasteurella multocida* toxin translocation domain. Mol Microbiol 54:239–250

Binz T, Rummel A (2009) Cell entry strategy of clostridial neurotoxins. J Neurochem 109: 1584–1595

Blocker D, Berod L, Fluhr JW, Orth J, Idzko M, Aktories K, Norgauer J (2006) *Pasteurella multocida* toxin (PMT) activates RhoGTPases, induces actin polymerization and inhibits migration of human dendritic cells, but does not influence macropinocytosis. Int Immunol 18:459–464

Blumenthal B, Hoffmann C, Aktories K, Backert S, Schmidt G (2007) The cytotoxic necrotizing factors from Yersinia pseudotuberculosis and from Escherichia coli bind to different cellular receptors but take the same route to the cytosol. Infect Immun 75:3344–3353

Boney CM, Fiedorek FT Jr, Paul SR, Gruppuso PA (1996) Regulation of preadipocyte factor-1 gene expression during 3T3-L1 cell differentiation. Endocrinology 137:2923–2928

Bourdon A, Minai L, Serre V, Jais JP, Sarzi E, Aubert S, Chretien D, de Lonlay P, Paquis-Flucklinger V, Arakawa H, Nakamura Y, Munnich A, Rotig A (2007) Mutation of RRM2B, encoding p53-controlled ribonucleotide reductase (p53R2), causes severe mitochondrial DNA depletion. Nat Genet 39:776–780

Brothers MC, Ho M, Maharjan R, Clemons NC, Bannai Y, Waites MA, Faulkner MJ, Kuhlenschmidt TB, Kuhlenschmidt MS, Blanke SR, Rienstra CM, Wilson BA (2011) Membrane interaction of Pasteurella multocida toxin involves sphingomyelin. FEBS J 278:4633–4648

Buetow L, Flatau G, Chiu K, Boquet P, Ghosh P (2001) Structure of the Rho-activating domain of *Escherichia coli* cytotoxic necrotizing factor 1. Nat Struct Biol 8:584–588

Bunemann M, Meyer T, Pott L, Hosey M (2000) Novel inhibition of gbetagamma-activated potassium currents induced by M(2) muscarinic receptors via a pertussis toxin-insensitive pathway. J Biol chem 275:12537–12545

Busch C, Orth J, Djouder N, Aktories K (2001) Biological activity of a C-terminal fragment of *Pasteurella multocida* toxin. Infect Immun 69:3628–3634

Buys WE, Smith HE, Kamps AM, Kamp EM, Smits MA (1990) Sequence of the dermonecrotic toxin of Pasteurella multocida ssp. multocida. Nucleic Acids Res 18:2815–2816

Chanter N, Rutter JM (1990) Colonisation by *Pasteurella multocida* in atrophic rhinitis of pigs and immunity to the osteolytic toxin. Vet Microbiol 25:253–265

Cheville NF, Rimler RB (1989) A protein toxin from *Pasteurella multocida* type D causes acute and chronic hepatic toxicity in rats. Vet Pathol 26:148–157

Cheville NF, Rimler RB, Thurston JR (1988) A toxin from *Pasteurella multocida* type D causes acute hepatic necrosis in pigs. Vet Pathol 25:518–520

Chrisp CE, Foged NT (1991) Induction of pneumonia in rabbits by use of a purified protein toxin from *Pasteurella multocida*. Am J Vet Res 52:56–61

Chung JW, Hong SJ, Kim KJ, Goti D, Stins MF, Shin S, Dawson VL, Dawson TM, Kim KS (2003) 37-kDa laminin receptor precursor modulates cytotoxic necrotizing factor 1-mediated RhoA activation and bacterial uptake. J Biol Chem 278:16857–16862. doi:10.1074/jbc.M301028200

Deeb BJ, DiGiacomo RF, Bernard BL, Silbernagel SM (1990) *Pasteurella multocida* and B*ordetella bronchiseptica* infections in rabbits. J Clin Microbiol 28:70–75

DiGiacomo RF, Allen V, Hinton MH (1991a) Naturally acquired *Pasteurella multocida* subsp. multocida infection in a closed colony of rabbits: characteristics of isolates. Lab Anim 25:236–241

DiGiacomo RF, Deeb BJ, Brodie SJ, Zimmerman TE, Veltkamp ER, Chrisp CE (1993) Toxin production by *Pasteurella multocida* isolated from rabbits with atrophic rhinitis. Am J Vet Res 54:1280–1286

DiGiacomo RF, Xu YM, Allen V, Hinton MH, Pearson GR (1991b) Naturally acquired *Pasteurella multocida* infection in rabbits: clinicopathological aspects. Can J Vet Res 55: 234–238

Dong M, Yeh F, Tepp WH, Dean C, Johnson EA, Janz R, Chapman ER (2006) SV2 is the protein receptor for botulinum neurotoxin A. Science 312:592–596

Donnio PY, Allardet-Servent A, Perrin M, Escande F, Avril JL (1999) Characterisation of dermonecrotic toxin-producing strains of Pasteurella multocida subsp. multocida isolated from man and swine. J Med Microbiol 48:125–131

Donnio PY, Avril JL, Andre PM, Vaucel J (1991) Dermonecrotic toxin production by strains of Pasteurella multocida isolated from man. J Med Microbiol 34:333–337

Donnio PY, Lerestif-Gautier AL, Avril JL (2004) Characterization of Pasteurella spp. strains isolated from human infections. J Comp Pathol 130:137–142

Dudet LI, Chailler P, Dubreuil JD, Martineau-Doize B (1996) *Pasteurella multocida* toxin stimulates mitogenesis and cytoskeleton reorganization in Swiss 3T3 fibroblasts. J Cell Physiol 168:173–182

Elling F, Pedersen KB, Hogh P, Foged NT (1988) Characterization of the dermal lesions induced by a purified protein from toxigenic Pasteurella multocida. APMIS 96:50–55

Essler M, Hermann K, Amano M, Kaibuchi K, Heesemann J, Weber PC, Aepfelbacher M (1998) *Pasteurella multocida* toxin increases endothelial permeability via Rho kinase and myosin light chain phosphatase. J Immunol 161:5640–5646

Falbo V, Pace T, Picci L, Pizzi E, Caprioli A (1993) Isolation and nucleotide sequence of the gene encoding cytotoxic necrotizing factor 1 of *Escherichia coli*. Infect Immun 61:4909–4914

Felix R, Fleisch H, Frandsen PL (1992) Effect of *Pasteurella multocida* toxin on bone resorption in vitro. Infect Immun 60:4984–4988

Flatau G, Lemichez E, Gauthier M, Chardin P, Paris S, Fiorentini C, Boquet P (1997) Toxin-induced activation of the G protein p21 Rho by deamidation of glutamine. Nature 387: 729–733

Foged NT (1992) *Pasteurella multocida* toxin. The characterisation of the toxin and its significance in the diagnosis and prevention of progressive atrophic rhinitis in pigs. APMIS Suppl 25:1–56

Foged NT, Nielsen JP, Jorsal SE (1989) Protection against progressive atrophic rhinitis by vaccination with *Pasteurella multocida* toxin purified by monoclonal antibodies. Vet Rec 125:7–11

Frederiksen W (1993) Ecology and significance of Pasteurellaceae in man–an update. Zentralbl Bakteriol 279:27–34

Frymus T, Bielecki W, Jakubowski T (1991) Toxigenic *Pasteurella multocida* in rabbits with naturally occurring atrophic rhinitis. Zentralbl Veterinarmed B 38:265–268

Frymus T, Muller E, Franz B, Petzoldt K (1989) Protection by toxoid-induced antibody of gnotobiotic piglets challenged with the dermonecrotic toxin of *Pasteurella multocida*. Zentralbl Veterinarmed B 36:674–680

Garces C, Ruiz-Hidalgo MJ, Bonvini E, Goldstein J, Laborda J (1999) Adipocyte differentiation is modulated by secreted delta-like (dlk) variants and requires the expression of membrane-associated dlk. Differentiation; research in biological diversity 64:103–114

Garcia VF (1997) Animal bites and *Pasturella* infections. Pediatr Rev 18:127–130

Geissler B, Tungekar R, Satchell KJ (2010) Identification of a conserved membrane localization domain within numerous large bacterial protein toxins. Proc Natl Acad Sci U S A 107:5581–5586

Griego RD, Rosen T, Orengo IF, Wolf JE (1995) Dog, cat, and human bites: a review. J Am Acad Dermatol 33:1019–1029

Gwaltney SM, Galvin RJ, Register KB, Rimler RB, Ackermann MR (1997) Effects of *Pasteurella multocida* toxin on porcine bone marrow cell differentiation into osteoclasts and osteoblasts. Vet Pathol 34:421–430

Hamilton TD, Roe JM, Hayes CM, Webster AJ (1998) Effect of ovalbumin aerosol exposure on colonization of the porcine upper airway by *Pasteurella multocida* and effect of colonization on subsequent immune function. Clin Diagn Lab Immunol 5:494–498

Harmey D, Stenbeck G, Nobes CD, Lax AJ, Grigoriadis AE (2004) Regulation of osteoblast differentiation by *Pasteurella multocida* toxin (PMT): a role for Rho GTPase in bone formation. J Bone Miner Res 19:661–670

Henderson SR, Shah A, Banford KB, Howard LS (2010) Pig trotters lung–novel domestic transmission of *Pasteurella multocida*. Clin Med 10:517–518

Hennig B, Orth J, Aktories K, Diener M (2008) Anion secretion evoked by *Pasteurella multocida* toxin across rat colon. Eur J Pharmacol 583:156–163

Hildebrand D, Heeg K, Kubatzky KF (2010a) *Pasteurella multocida* toxin-stimulated osteoclast differentiation is B cell dependent. Infect Immun 79:220–228

Hildebrand D, Walker P, Dalpke A, Heeg K, Kubatzky KF (2010b) *Pasteurella multocida* toxin-induced Pim-1 expression disrupts suppressor of cytokine signalling (SOCS)-1 activity. Cell Microbiol 12:1732–1745

Hoffmann C, Schmidt G (2004) CNF and DNT. Rev Physiol Biochem Pharmacol 152:49–63

Holst E, Rollof J, Larsson L, Nielsen JP (1992) Characterization and distribution of Pasteurella species recovered from infected humans. J Clin Microbiol 30:2984–2987

Horiguchi Y (2001) *Escherichia coli* cytotoxic necrotizing factors and *Bordetella* dermonecrotic toxin: the dermonecrosis-inducing toxins activating Rho small GTPases. Toxicon 39:1619–1627

Hoskins IC, Thomas LH, Lax AJ (1997) Nasal infection with *Pasteurella multocida* causes proliferation of bladder epithelium in gnotobiotic pigs. Vet Rec 140:22

Iaria C, Cascio A (2007) Please, do not forget *Pasteurella multocida*. Clin Infect Dis 45:940

Jordan RW, Hamilton TD, Hayes CM, Patel D, Jones PH, Roe JM, Williams NA (2003) Modulation of the humoral immune response of swine and mice mediated by toxigenic *Pasteurella multocida*. FEMS Immunol Med Microbiol 39:51–59

Jutras I, Martineau-Doize B (1996) Stimulation of osteoclast-like cell formation by *Pasteurella multocida* toxin from hemopoietic progenitor cells in mouse bone marrow cultures. Can J Vet Res 60:34–39

Kamps AM, Kamp EM, Smits MA (1990) Cloning and expression of the dermonecrotic toxin gene of Pasteurella multocida ssp. multocida in Escherichia coli. FEMS Microbiol Lett 55:187–190

Katada T, Oinuma M, Ui M (1986) Two guanine nucleotide-binding proteins in rat brain serving as the specific substrate of islet-activating protein, pertussis toxin. Interaction of the alpha subunits with beta gamma-subunits in development of their biological activities. J Biol Chem 261:8182–8191

Kielstein P (1986) On the occurrence of toxin-producing *Pasteurella multocida* strains in atrophic rhinitis and in pneumonias of swine and cattle. Zentralbl Veterinarmed B 33:418–424

Kim KJ, Chung JW, Kim KS (2005) 67-kDa laminin receptor promotes internalization of cytotoxic necrotizing factor 1-expressing Escherichia coli K1 into human brain microvascular endothelial cells. J Biol Chem 280:1360–1368

Kimman TG, Kamp EM (1986) Induced atrophic rhinitis in rats. Am J Vet Res 47:2426–2430

Kimman TG, Lowik CW, van de Wee-Pals LJ, Thesingh CW, Defize P, Kamp EM, Bijvoet OL (1987) Stimulation of bone resorption by inflamed nasal mucosa, dermonecrotic toxin-containing conditioned medium from *Pasteurella multocida*, and purified dermonecrotic toxin from *P. multocida*. Infect Immun 55:2110–2116

Kitadokoro K, Kamitani S, Miyazawa M, Hanajima-Ozawa M, Fukui A, Miyake M, Horiguchi Y (2007) Crystal structures reveal a thiol protease-like catalytic triad in the C-terminal region of *Pasteurella multocida* toxin. Proc Natl Acad Sci U S A 104:5139–5144

Klein NC, Cunha BA (1997) *Pasteurella multocida* pneumonia. Semin Respir Infect 12:54–56

Kobayaa H, Souki RR, Trust S, Domachowske JB (2009) *Pasteurella multocida* meningitis in newborns after incidental animal exposure. Pediatr Infect Dis J 28:928–929

Lacerda HM, Lax AJ, Rozengurt E (1996) *Pasteurella multocida* toxin, a potent intracellularly acting mitogen, induces p125FAK and paxillin tyrosine phosphorylation, actin stress fiber formation, and focal contact assembly in Swiss 3T3 cells. J Biol Chem 271:439–445

Lax AJ, Chanter N (1990) Cloning of the toxin gene from *Pasteurella multocida* and its role in atrophic rhinitis. J Gen Microbiol 136:81–87

Lemichez E, Flatau G, Bruzzone M, Boquet P, Gauthier M (1997) Molecular localization of the *Escherichia coli* cytotoxic necrotizing factor CNF1 cell-binding and catalytic domains. Mol Microbiol 24:1061–1070

Liu L, Clipstone NA (2007) Prostaglandin F2alpha inhibits adipocyte differentiation via a G alpha q-calcium-calcineurin-dependent signaling pathway. J Cell Biochem 100:161–173

Lockman HA, Gillespie RA, Baker BD, Shakhnovich E (2002) Yersinia pseudotuberculosis produces a cytotoxic necrotizing factor. Infect Immun 70:2708–2714

Luo S, Ho M, Wilson BA (2008) Application of intact cell-based NFAT-beta-lactamase reporter assay for *Pasteurella multocida* toxin-mediated activation of calcium signaling pathway. Toxicon 51:597–605

Magyar T (1989) Study of the toxin-producing ability of *Pasteurella multocida* in mice. Acta Vet Hung 37:319–325

Martineau-Doize B, Caya I, Gagne S, Jutras I, Dumas G (1993) Effects of *Pasteurella multocida* toxin on the osteoclast population of the rat. J Comp Pathol 108:81–91

Merritt EA, Sarfaty S, van den Akker F, L'Hoir C, Martial JA, Hol WG (1994) Crystal structure of cholera toxin B-pentamer bound to receptor GM1 pentasaccharide. Protein Sci 3:166–175

Meyer T, Wellner-Kienitz MC, Biewald A, Bender K, Eickel A, Pott L (2001) Depletion of phosphatidylinositol 4,5-bisphosphate by activation of phospholipase C-coupled receptors causes slow inhibition but not desensitization of G protein-gated inward rectifier K+ current in atrial myocytes. J Biol Chem 276:5650–5658

Migliore E, Serraino C, Brignone C, Ferrigno D, Cardellicchio A, Pomero F, Castagna E, Osenda M, Fenoglio L (2009) *Pasteurella multocida* infection in a cirrhotic patient: case report, microbiological aspects and a review of literature. Adv Med Sci 54:109–112

Mullan PB, Lax AJ (1996) *Pasteurella multocida* toxin is a mitogen for bone cells in primary culture. Infect Immun 64:959–965

Mullan PB, Lax AJ (1998) *Pasteurella multocida* toxin stimulates bone resorption by osteoclasts via interaction with osteoblasts. Calcif Tissue Int 63:340–345

Murphy AC, Rozengurt E (1992) *Pasteurella multocida* toxin selectively facilitates phosphatidylinositol 4,5-bisphosphate hydrolysis by bombesin, vasopressin, and endothelin. Requirement for a functional G protein. J Biol Chem 267:25296–25303

Naglich JG, Metherall JE, Russell DW, Eidels L (1992) Expression cloning of a diphtheria toxin receptor: identity with a heparin-binding EGF-like growth factor precursor. Cell 69:1051–1061

Nakai T, Sawata A, Tsuji M, Kume K (1984) Characterization of dermonecrotic toxin produced by serotype D strains of *Pasteurella multocida*. Am J Vet Res 45:2410–2413

Neal JW, Clipstone NA (2002) Calcineurin mediates the calcium-dependent inhibition of adipocyte differentiation in 3T3-L1 cells. J Biol Chem 277:49776–49781. doi:10.1074/jbc.M207913200M207913200

Obreztchikova M, Elouardighi H, Ho M, Wilson BA, Gertsberg Z, Steinberg SF (2006) Distinct signaling functions for Shc isoforms in the heart. J Biol Chem 281:20197–20204

Ohnishi T, Horiguchi Y, Masuda M, Sugimoto N, Matsuda M (1998) *Pasteurella multocida* toxin and *Bordetella bronchiseptica* dermonecrotizing toxin elicit similar effects on cultured cells by different mechanisms. J Vet Med Sci 60:301–305

Orth JH, Aktories K, Kubatzky KF (2007a) Modulation of host cell gene expression through activation of STAT transcription factors by *Pasteurella multocida* toxin. J Biol Chem 282: 3050 3057

Orth JH, Blocker D, Aktories K (2003) His1205 and His1223 are essential for the activity of the mitogenic *Pasteurella multocida* toxin. Biochemistry 42:4971–4977

Orth JH, Fester I, Preuss I, Agnoletto L, Wilson BA, Aktories K (2008) Activation of Galpha (i) and subsequent uncoupling of receptor-Galpha(i) signaling by *Pasteurella multocida* toxin. J Biol Chem 283:23288–23294

Orth JH, Lang S, Aktories K (2004) Action of *Pasteurella multocida* toxin depends on the helical domain of Galphaq. J Biol Chem 279:34150–34155

Orth JH, Lang S, Preuss I, Milligan G, Aktories K (2007b) Action of *Pasteurella multocida* toxin on Galpha(q) is persistent and independent of interaction with G-protein-coupled receptors. Cell Signal 19:2174–2182

Orth JH, Lang S, Taniguchi M, Aktories K (2005) *Pasteurella multocida* toxin-induced activation of RhoA is mediated via two families of G{alpha} proteins, G{alpha}q and G{alpha}12/13. J Biol Chem 280:36701–36707

Orth JH, Preuss I, Fester I, Schlosser A, Wilson BA, Aktories K (2009) *Pasteurella multocida* toxin activation of heterotrimeric G proteins by deamidation. Proc Natl Acad Sci U S A 106:7179–7184

Oswald E, Sugai M, Labigne A, Wu HC, Fiorentini C, Boquet P, O'Brien AD (1994) Cytotoxic necrotizing factor type 2 produced by virulent *Escherichia coli* modifies the small GTP-binding proteins Rho involved in assembly of actin stress fibers. Proc Natl Acad Sci U S A 91:3814–3818

Ozgen N, Obreztchikova M, Guo J, Elouardighi H, Dorn GW 2nd, Wilson BA, Steinberg SF (2008) Protein kinase D links Gq-coupled receptors to cAMP response element-binding protein (CREB)-Ser133 phosphorylation in the heart. J Biol Chem 283:17009–17019

Pei S, Doye A, Boquet P (2001) Mutation of specific acidic residues of the CNF1 T domain into lysine alters cell membrane translocation of the toxin. Mol Microbiol 41:1237–1247

Pennings AM, Storm PK (1984) A test in vero cell monolayers for toxin production by strains of *Pasteurella multocida* isolated from pigs suspected of having atrophic rhinitis. Vet Microbiol 9:503–8. doi:0378-1135(84)90071-3

Petersen SK, Foged NT (1989) Cloning and expression of the Pasteurella multocida toxin gene, toxA, in Escherichia coli. Infect Immun 57:3907–3913

Pettit RK, Ackermann MR, Rimler RB (1993a) Receptor-mediated binding of *Pasteurella multocida* dermonecrotic toxin to canine osteosarcoma and monkey kidney (vero) cells. Lab Invest 69:94–100

Pettit RK, Rimler RB, Ackermann MR (1993b) Protection of *Pasteurella multocida* dermonecrotic toxin-challenged rats by toxoid-induced antibody. Vet Microbiol 34:167–173

Preuss I, Hildebrand D, Orth JH, Aktories K, Kubatzky KF (2010) *Pasteurella multocida* toxin is a potent activator of anti-apoptotic signalling pathways. Cell Microbiol 12:1174–1185

Preuss I, Kurig B, Nurnberg B, Orth JH, Aktories K (2009) *Pasteurella multocida* toxin activates Gbetagamma dimers of heterotrimeric G proteins. Cell Signal 21:551–558

Pullinger GD, Adams TE, Mullan PB, Garrod TI, Lax AJ (1996) Cloning, expression, and molecular characterization of the dermonecrotic toxin gene of Bordetella spp. Infect Immun 64:4163–4171

Pullinger GD, Lax AJ (2007) Histidine residues at the active site of the Pasteurella multocida toxin. Open Biochem J 1:7–11

Pullinger GD, Sowdhamini R, Lax AJ (2001) Localization of functional domains of the mitogenic toxin of *Pasteurella multocida*. Infect Immun 69:7839–7850

Repella TL, Ho M, Chong TP, Bannai Y, Wilson BA (2011) Arf6-dependent intracellular trafficking of pasteurella multocida toxin and pH-dependent translocation from late endosomes. Toxins 3:218–241

Rozengurt E, Higgins T, Chanter N, Lax AJ, Staddon JM (1990) *Pasteurella multocida* toxin: potent mitogen for cultured fibroblasts. Proc Natl Acad Sci U S A 87:123–127

Rutter JM, Luther PD (1984) Cell culture assay for toxigenic *Pasteurella multocida* from atrophic rhinitis of pigs. Vet Rec 114:393–396

Sabri A, Pak E, Alcott SA, Wilson BA, Steinberg SF (2000) Coupling function of endogenous alpha(1)- and beta-adrenergic receptors in mouse cardiomyocytes. Circ Res 86:1047–1053

Sabri A, Wilson BA, Steinberg SF (2002) Dual actions of the Galpha(q) agonist *Pasteurella multocida* toxin to promote cardiomyocyte hypertrophy and enhance apoptosis susceptibility. Circ Res 90:850–857

Sagi SA, Seasholtz TM, Kobiashvili M, Wilson BA, Toksoz D, Brown JH (2001) Physical and functional interactions of Galphaq with Rho and its exchange factors. J Biol Chem 276:15445–15452

Satomura A, Yanai M, Fujita T, Arashima Y, Kumasaka K, Nakane C, Ito K, Fuke Y, Maruyama T, Maruyama N, Okada K, Nakayama T, Matsumoto K (2010) Peritonitis associated with Pasteurella multocida: molecular evidence of zoonotic etiology. Ther Apher Dial 14:373–376

Schmidt G, Sehr P, Wilm M, Selzer J, Mann M, Aktories K (1997) Gln 63 of Rho is deamidated by *Escherichia coli* cytotoxic necrotizing factor-1. Nature 387:725–729

Schmidt G, Selzer J, Lerm M, Aktories K (1998) The Rho-deamidating cytotoxic necrotizing factor 1 from *Escherichia coli* possesses transglutaminase activity. Cysteine 866 and histidine 881 are essential for enzyme activity. J Biol Chem 273:13669–13674

Seo B, Choy EW, Maudsley S, Miller WE, Wilson BA, Luttrell LM (2000) *Pasteurella multocida* toxin stimulates mitogen-activated protein kinase via G(q/11)-dependent transactivation of the epidermal growth factor receptor. J Biol Chem 275:2239–2245

Staddon JM, Barker CJ, Murphy AC, Chanter N, Lax AJ, Michell RH, Rozengurt E (1991) *Pasteurella multocida* toxin, a potent mitogen, increases inositol 1,4,5-trisphosphate and mobilizes Ca^{2+} in Swiss 3T3 cells. J Biol Chem 266:4840–4847

Staddon JM, Bouzyk MM, Rozengurt E (1992) Interconversion of GRP78/BiP. A novel event in the action of *Pasteurella multocida* toxin, bombesin, and platelet-derived growth factor. J Biol Chem 267:25239–25245

Staddon JM, Chanter N, Lax AJ, Higgins TE, Rozengurt E (1990) *Pasteurella multocida* toxin, a potent mitogen, stimulates protein kinase C-dependent and -independent protein phosphorylation in Swiss 3T3 cells. J Biol Chem 265:11841–11848

Stein PE, Boodhoo A, Armstrong GD, Heerze LD, Cockle SA, Klein MH, Read RJ (1994) Structure of a pertussis toxin-sugar complex as a model for receptor binding. Nat Struct Biol 1:591–596

Sterner-Kock A, Lanske B, Uberschar S, Atkinson MJ (1995) Effects of the *Pasteurella multocida* toxin on osteoblastic cells in vitro. Vet Pathol 32:274–279

Stoll T, Markwirth G, Reipschlager S, Schmidt G (2009) A new member of a growing toxin family–Escherichia coli cytotoxic necrotizing factor 3 (CNF3). Toxicon 54:745–753

Suckow MA (2000) Immunization of rabbits against *Pasteurella multocida* using a commercial swine vaccine. Lab Anim 34:403–408

Suckow MA, Bowersock TL, Nielsen K, Chrisp CE, Frandsen PL, Janovitz EB (1995) Protective immunity to *Pasteurella multocida* heat-labile toxin by intranasal immunization in rabbits. Lab Anim Sci 45:526–532

Sul HS (2009) Minireview: Pref-1: role in adipogenesis and mesenchymal cell fate. Mol Endocrinol 23:1717–1725

Thomas W, Pullinger GD, Lax AJ, Rozengurt E (2001) *Escherichia coli* cytotoxic necrotizing factor and *Pasteurella multocida* toxin induce focal adhesion kinase autophosphorylation and Src association. Infect Immun 69:5931–5935

Thurston JR, Rimler RB, Ackermann MR, Cheville NF (1992) Use of rats to compare atrophic rhinitis vaccines for protection against effects of heat-labile protein toxin produced by *Pasteurella multocida* serogroup D. Vet Immunol Immunopathol 33:155–162

van Diemen PM, de Vries Reilingh G, Parmentier HK (1994) Immune responses of piglets to *Pasteurella multocida* toxin and toxoid. Vet Immunol Immunopathol 41:307–321

van Diemen PM, de Vries Reilingh G, Parmentier HK (1996) Effect of *Pasteurella multocida* toxin on in vivo immune responses in piglets. Vet Q 18:141–146

Waldor M, Roberts D, Kazanjian P (1992) In utero infection due to *Pasteurella multocida* in the first trimester of pregnancy: case report and review. Clin Infect Dis 14:497–500

Wilson BA, Aminova LR, Ponferrada VG, Ho M (2000) Differential modulation and subsequent blockade of mitogenic signaling and cell cycle progression by *Pasteurella multocida* toxin. Infect Immun 68:4531–4538

Wilson BA, Ho M (2006) *Pasteurella multocida* toxin. In: Alouf JE, Popoff MR (eds) The Comprehensive Sourcebook of Bacterial Protein Toxins. Elsevier Science Publishers B. V, Amsterdam, pp 430–447

Wilson BA, Ho M (2010) Recent insights into *Pasteurella multocida* toxin and other G-protein-modulating bacterial toxins. Future Microbiol 5:1185–1201

Wilson BA, Ho M (2011) Cellular and molecular action of the mitogenic protein-deamidating toxin from *Pasteurella multocida*. FEBS J 278:4616–4632

Wilson BA, Zhu X, Ho M, Lu L (1997) *Pasteurella multocida* toxin activates the inositol triphosphate signaling pathway in *Xenopus* oocytes via G(q)alpha-coupled phospholipase C-beta1. J Biol Chem 272:1268–1275

Zywietz A, Gohla A, Schmelz M, Schultz G, Offermanns S (2001) Pleiotropic effects of *Pasteurella multocida* toxin are mediated by Gq-dependent and -independent mechanisms. Involvement of Gq but not G11. J Biol Chem 276:3840–3845

Swine Atrophic Rhinitis Caused by *Pasteurella multocida* Toxin and *Bordetella* Dermonecrotic Toxin

Yasuhiko Horiguchi

Abstract Atrophic rhinitis is a widespread and economically important swine disease caused by *Pasteurella multocida* and *Bordetella bronchiseptica*. The disease is characterized by atrophy of the nasal turbinate bones, which results in a shortened and deformed snout in severe cases. *P. multocida* toxin and *B. bronchiseptica* dermonecrotic toxin have been considered to independently or cooperatively disturb the osteogenesis of the turbinate bone by inhibiting osteoblastic differentiation and/or stimulating bone resorption by osteoclasts. Recently, the intracellular targets and molecular actions of both toxins have been clarified, enabling speculation on the intracellular signals leading to the inhibition of osteogenesis.

Contents

Y. Horiguchi (✉)
Department of Molecular Bacteriology,
Research Institute for Microbial Diseases,
Osaka University, Yamada-oka 3-1, Suita, Osaka 565-0871, Japan
e-mail: horiguti@biken.osaka-u.ac.jp

Current Topics in Microbiology and Immunology (2012) 361: 113–129
DOI: 10.1007/82_2012_206

Published Online: 13 March 2012

1 Atrophic Rhinitis

Atrophic rhinitis (AR) is an upper respiratory disease of pigs. The clinical signs of AR include sneezing, coughing, nasal discharge, and epistaxis. Atrophy of the turbinate bones, which results in deformation of the snout, and dark crescent-shaped stains below the eyes are also characteristic pathological changes in infected pigs. Diseased piglets often show growth retardation, which causes economic loss to the pig rearing industry. Since the disease was first reported in the early nineteenth century, its etiology has been controversial, with researchers suggesting the involvement of a heritable factor, nutritional conditions, and bacterial or viral infections (Rutter 1985). However, it is now understood that *Bordetella bronchiseptica* and *Pasteurella multocida* are the causative agents for AR. Notably, it was considered that *B. bronchiseptica* causes moderate turbinate lesions whereas *P. multocida*, which by itself poorly colonizes the nasal cavity, establishes a secondary infection only after *B. bronchiseptica*, and exacerbates the disease (Harris and Switzer 1968; Pedersen and Barfod 1981; Pedersen and Elling 1984; Rutter 1985). The disease exacerbated by the secondary infection by *P. multocida* is often called "progressive AR" to distinguish it from "nonprogressive AR" caused by the single infection of *B. bronchiseptica*. These days, *P. multocida* is also recognized as the primary pathogen causing progressive AR without other agents like *B. bronchiseptica* (de Jong and Nielsen 1990). For details of early studies on the clinical and pathological features of AR in relation to the pathogenesis of *B. bronchiseptica* and *P. multocida*, the reader is referred to a previous review article (Rutter 1985).

Atrophy of the turbinate bones is the most characteristic change in AR. Histopathological analyses have revealed both degenerative changes of osteoblasts and osteocytes and bone resorption. In 1979, Hanada et al. succeeded in producing turbinate atrophy in piglets via the intranasal inoculation of extracts from *B. bronchiseptica* that produced high levels of dermonecrotic toxin (DNT), formerly called heat-labile toxin (Hanada et al. 1979). In addition, turbinate atrophy was induced by the administration of sterile culture fluid of toxigenic *P. multocida* (Pedersen and Barfod 1981; de Jong 1983; Rutter and Mackenzie 1984; Pedersen and Elling 1984). Therefore, toxic substances from *B. bronchiseptica* and *P. multocida* were considered responsible for AR, or at least the bone lesions. These toxins, which are currently called DNT and *P. multocida* toxin (PMT), were purified and characterized in 1989 and 1984, respectively (Nakai et al. 1984b;

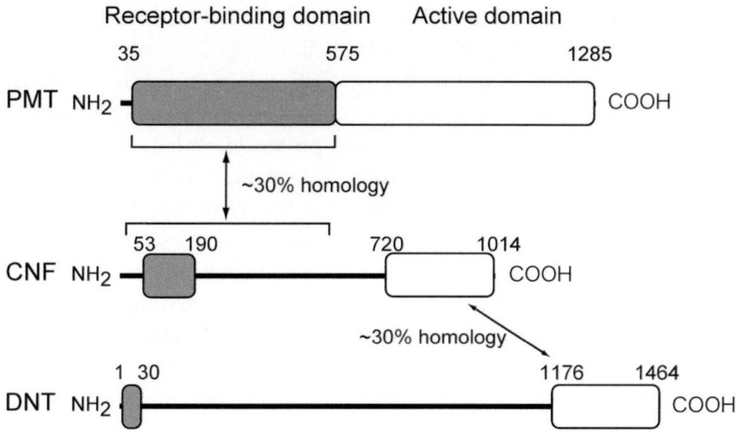

Fig. 1 Locations of functional domains of the dermonecrosis-inducing toxins. All the toxins carry the N-terminal receptor-binding domain (*gray box*) and the C-terminal active domain (*open box*). The receptor-binding domain of PMT is not well-defined and includes translocation domain spanning amino-acid residues 402–457 (Baldwin et al. 2004)

Horiguchi et al. 1989). Eventually, it was reported that the purified DNT and PMT produced bone lesions similar to those seen in AR (Dominick and Rimler 1986, 1988; Kimman et al. 1987; Horiguchi et al. 1995a). Thus, these toxins are now accepted as causative agents for this disease. Because previous studies pointed out that *B. bronchiseptica* impaired osteoblastic activity, whereas *P. multocida* stimulated osteoclastic activity (Fetter et al. 1975; Silveira et al. 1982; Pedersen and Elling 1984; Kimman et al. 1987), the toxic actions of DNT and PMT leading to bone atrophy could be distinct, as discussed later.

2 The Toxins Causing the Pathological Changes in Bone Tissues

2.1 DNT

DNT is commonly produced by organisms belonging to the genus *Bordetella,* such as *B. bronchiseptica*, *B. parapertussis*, and *B. pertussis*. The name DNT comes from the characteristic skin lesion that the toxin produces when injected intradermally into animals. This toxin is a single-chain polypeptide, which consists of 1,464 amino acids (Kashimoto et al. 1999). The N-terminal region containing the first 30 amino acids is involved in binding to target cells, whereas the C-terminal 300 amino acids comprises the enzymatically active domain, which shows homology with the C-terminal region of cytotoxic necrotizing factors (CNFs) produced by *Escherichia coli* or *Yersinia pseudotuberculosis* (Fig. 1) (Walker and

Weiss 1994; Lemichez et al. 1997; Kashimoto et al. 1999; Matsuzawa et al. 2002; Hoffmann and Schmidt 2004; Knust and Schmidt 2010; Fukui-Miyazaki et al. 2011). DNT and CNFs exert similar but distinct enzymatic activities as described below. The region encompassing amino acids 45–166 includes a putative transmembrane domain and the toxin undergoes proteolytic cleavage between Arg44 and Glu45 by a furin-like protease (Matsuzawa et al. 2004). Thus, it is currently considered that DNT binds to target cells via the N-terminal region, undergoes intramolecular cleavage, and translocates the enzymatically active region into the cytoplasm (Fukui and Horiguchi 2004; Fukui-Miyazaki et al. 2011). The enzymatically active domain possesses transglutaminase activity that specifically deamidates or polyaminates Rho family GTPases (Horiguchi et al. 1997; Masuda et al. 2000). The Rho GTPases become constitutively active, stimulating downstream signaling pathways, which affect various cellular functions.

2.2 PMT

PMT is produced by toxigenic *P. multocida* which harbors *toxA*, the PMT gene, on a lysogenic bacteriophage (Nagai et al. 1994; Pullinger et al. 2004). Once it was considered that PMT might at least partly share homology with DNT, because these toxins show similar biological activities; both toxins cause dermonecrosis in guinea pigs on intradermal injection, and atrophy of the spleen and death in mice. However, anti-PMT serum did not neutralize the toxicity of DNT and vice versa, indicating these two toxins to be serologically distinct (Nakai et al. 1984a). Subsequently, the amino-acid sequences deduced from the gene sequences clearly showed that these toxins are quite distinct (Petersen 1990; Lax et al. 1990). PMT is a single-chain polypeptide, which comprises 1,285 amino acids (Buys et al. 1990; Petersen 1990; Lax et al. 1990). The receptor-binding domain of PMT is believed to reside in the N-terminal region, which shows homology to those of CNFs (Fig. 1) (Oswald et al. 1994; Walker and Weiss 1994; Lemichez et al. 1997; Busch et al. 2001; Pullinger et al. 2001; Baldwin et al. 2004). According to the crystal structure (Fig. 2), the C-terminal part of ~700 amino acids comprises the intracellular active unit, which is composed of three domains named C1, C2, and C3 from the N terminus to the C terminus (Busch et al. 2001; Pullinger et al. 2001; Kitadokoro et al. 2007). The C3 domain contains the enzyme moiety that deamidates some types of α subunits of heterotrimeric GTPases such as G_i and G_q (Kitadokoro et al. 2007; Aminova et al. 2008; Orth et al. 2009). C1 functions to recruit the toxin to the inner leaflet of the plasma membrane where the target Gα subunits reside (Kamitani et al. 2010). The role of C2 in the toxin's actions remains unknown. The Gα subunits deamidated by PMT become constitutively active and transduce positive signals downstream, like the Rho GTPases deamidated by DNT. The details of PMT actions are discussed below.

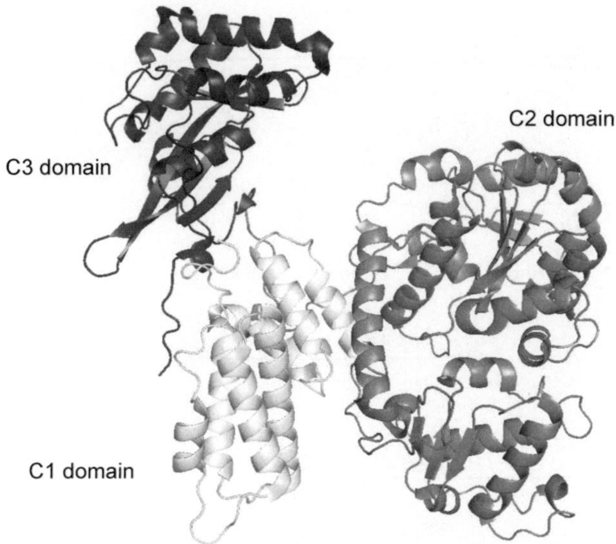

Fig. 2 The crystal structure of the C-terminal active domain of PMT. The active domain can be divided into three subdomains, C1, C2, and C3, which have distinct roles as described in the text

2.3 Dermonecrosis-Inducing Toxin Family

PMT, DNT, and CNFs are often referred to as members of the dermonecrosis-inducing toxin family (Fig. 1). All these toxins are single-chain polypeptides with receptor-binding domains located at the N-terminus, and the biologically active domain (catalytic domains) at the C-terminus (Oswald et al. 1994; Walker and Weiss 1994; Lemichez et al. 1997; Busch et al. 2001; Pullinger et al. 2001; Baldwin et al. 2004; Ward et al. 1998; Kashimoto et al. 1999; Matsuzawa et al. 2002; Orth et al. 2003; Fukui-Miyazaki et al. 2011). The catalytic domain of DNT shows ~30% similarity with those of CNFs produced by *E. coli* or *Y. pseudotuberculosis*, which are located at the C-terminal regions. On the other hand, the receptor-binding domain of PMT shares ~30% similarity with those of CNFs at the N-terminus (Oswald et al. 1994; Walker and Weiss 1994; Lemichez et al. 1997; Kashimoto et al. 1999; Hoffmann and Schmidt 2004). When injected intradermally, these toxins cause a characteristic dermonecrosis in guinea pigs, rabbits, and mice (Caprioli et al. 1983; Nakai et al. 1984a). Among them, DNT and CNFs activate the Rho family GTPases, whereas PMT targets α-subunits of the heterotrimeric GTPases; PMT and DNT but not CNFs are involved in the turbinate atrophy in AR.

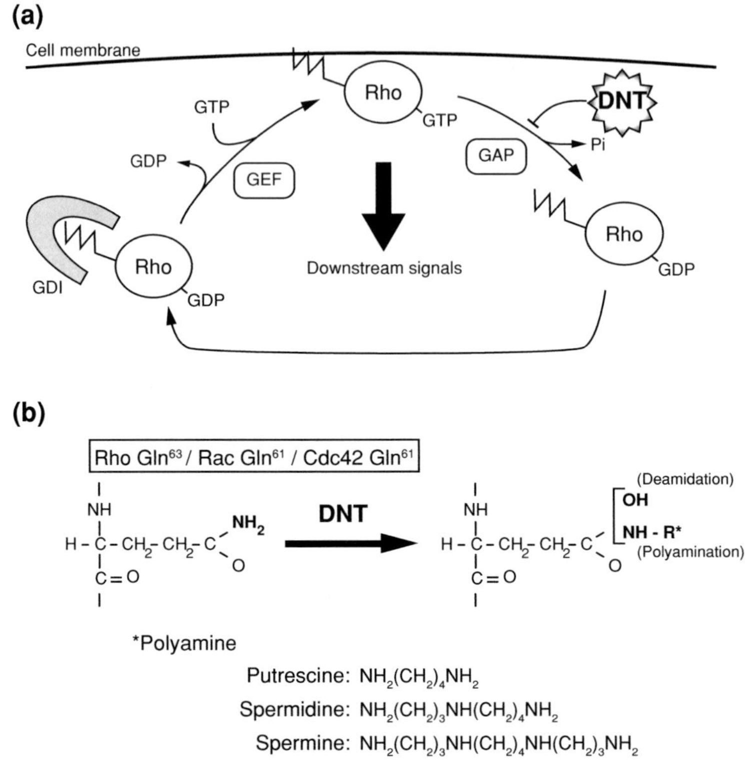

Fig. 3 **a** Regulation of the small GTPase as a molecular switch. The figure represents Rho, which is sensitive to DNT and implicated in osteoblastic differentiation. *GDI* GDP dissociation inhibitor, *GEF* GDP/GTP exchange factor, *GAP* GTPase-activating protein. **b** Deamidation, and polyamination of Rho GTPases by DNT

3 Intracellular Signaling Pathways Influenced by DNT and PMT

3.1 Molecular Actions of DNT on Rho GTPases

As mentioned above, the Rho GTPases including Rho, Rac, and Cdc42 are targets for DNT. The small GTPases are known to function as molecular switches of signal transduction pathways conducting various processes in eucaryotic cells (Fig. 3). They exist in two interconvertible forms: an inactive GDP-bound form and an active GTP-bound form. The GDP-bound GTPase in resting cells exchanges GDP for GTP on various types of stimulation, transduces signals to downstream effector proteins, and thereafter, returns to the inactive GDP-bound form by hydrolyzing the bound GTP. Many accessory factors modulate these processes: The GDP–GTP exchange reaction is inhibited by GDP dissociation inhibitor, but accelerated by GDP/GTP exchange factor. The intrinsic GTPase activity is activated by GTPase-activating protein (GAP).

DNT deamidates or polyaminates the Gln63 of Rho and the corresponding Gln61 of Rac and Cdc42 (Fig. 3) (Horiguchi et al. 1997; Masuda et al. 2000). The deamidation and the polyamination are essentially the same acyl transfer reaction, in which the γ-carboxamide group of Gln is an acyl donor. In the case of the DNT-catalyzing reaction, primary amino groups in ubiquitous polyamines including putrescine, spermidine, and spermine function as acyl acceptors, and in the absence or at less than threshold concentrations of polyamines, water acts as an acyl acceptor to accomplish deamidation. DNT seems to preferentially catalyze polyamination rather than deamidation (Masuda et al. 2000).

The glutamines that are modified by DNT reside in the switch II region responsible for the GTP-hydrolyzing activity of the GTPases. After being modified, the GTPases lose both the intrinsic and the GAP-stimulated GTPase activities but retain the GTP-binding activity (Horiguchi et al. 1997) (Fig. 3). Therefore, the modified GTPases become constitutively active and upregulate the downstream signal transductions. In addition, the polyaminated Rho acquires the ability to associate in a GTP-independent manner with the downstream effector ROCK (Masuda et al. 2000). Thus, in cells intoxicated by DNT, almost all of the signaling pathways that are dependent on the Rho family GTPases are constitutively stimulated. In fact, cultured cells intoxicated with DNT showed an aberrant organization of the actin cytoskeleton, including stress fibers, lamellipodia, and filopodia, which is known to result from the activation of Rho GTPases (Horiguchi et al. 1995b, 1997; Matsuzawa et al. 2002).

3.2 Molecular Actions of PMT on the Heterotrimeric GTPases

It has long been reported that PMT stimulates signaling pathways dependent on heterotrimeric GTPases such as G_q, $G_{12/13}$ and G_i (Murphy and Rozengurt 1992; Zywietz et al. 2001; Orth et al. 2005). The heterotrimeric GTPases transduce signals raised by the binding of ligands to G-protein-coupled receptors (GPCRs) to the downstream pathways (Fig. 4). In a resting state without a ligand, the GDP-bound α-subunit is associated with $\beta\gamma$-subunits. Once a ligand binds to the GPCR, the α-subunit exchanges GDP for GTP and releases the $\beta\gamma$-subunits. The α- and $\beta\gamma$-subunits independently transduce signals downstream. The bound GTP is hydrolyzed by the intrinsic or regulators of G-protein signaling (RGS)-stimulated GTP-hydrolyzing activity, and in turn, the GTPases again enter a GDP-bound state. Recently, Orth et al. reported that PMT specifically deamidates the catalytic residues of the α-subunits (e.g., Gln205 for $G\alpha_{i2}$ and Gln209 for $G\alpha_q$) and locks the GTPases in an active state by abolishing their GTP-hydrolyzing activity (Fig. 4, also see Fig. 3 for deamidation of Gln) (Orth et al. 2009). Later, deamidation of $G\alpha_{11}$ was also demonstrated (Kamitani et al. 2011). The α-subunits of G_{12} and G_{13} should be also deamidated by the toxin, because the downstream

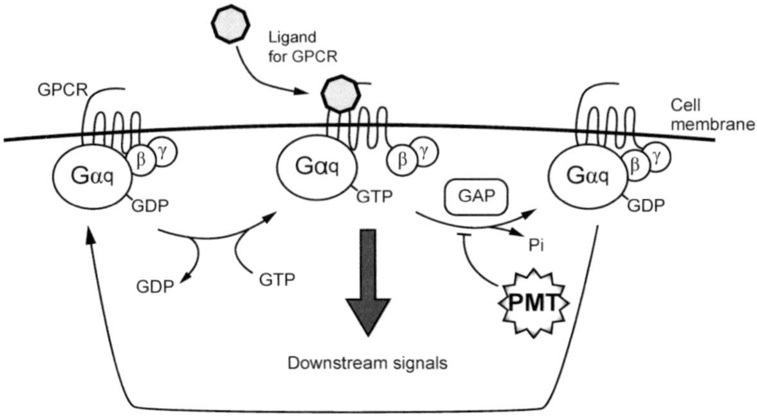

Fig. 4 Regulation of the heterotrimeric GTPase. In this figure, G_q, which is a target for PMT, is represented. *GPCR* G-protein-coupled receptor. *GAP* GTPase-activating protein. In the regulatory cycle of the heterotrimeric GTPase, the $\beta\gamma$-subunits, RGS, and GPCR serve as GDI, GAP, and GEF for the α-subunit, respectively

signals are stimulated by PMT. The constitutive activation of G_q leads to the activation of phospholipase Cβ, which hydrolyzes phosphatidylinositol 4,5-bisphosphate to generate inositol 1,4,5-triphosphate (IP3) and diacylglycerol (DG). IP3 and DG in turn lead to an increase in the intracellular Ca^{2+} and the activation of protein kinase C (PKC) (Staddon et al. 1990; Murphy and Rozengurt 1992). These signaling pathways might be important for atrophy of the turbinate bone as mentioned below. Rho GTPase is also known to be activated by PMT, which is probably mediated by the activation of G_q and $G_{12/13}$ (Orth et al. 2005). The actions of the toxin on G_i result in the inhibition of adenylyl cyclase and a reduction in cAMP accumulation stimulated by forskolin (Orth et al. 2008, 2009).

4 Cellular Actions of DNT and PMT

4.1 General Aspects

At a cellular level, DNT alters cell morphology (Horiguchi et al. 1991), stimulates DNA and protein syntheses (Horiguchi et al. 1993, 1994), and inhibits osteoblastic differentiation (Horiguchi et al. 1991). The DNT-induced morphological changes in target cells are accompanied by the formation of actin stress fibers and focal adhesions, which is regulated by the small GTPase Rho (Horiguchi et al. 1995b). Multinucleation in DNT-treated cells is often seen. The cell number does not increase after treatment with DNT, despite the fact that DNA synthesis is stimulated, indicating that the multinucleation occurs probably as a result of interference with cytokinesis (Horiguchi et al. 1993). Receptor molecules for DNT have not been identified. Only a few established lines of cultured cells such as MC3T3-E1, swiss3T3, and C3H10T1/2 are sensitive to DNT (Horiguchi et al. 1991, 1995b; Matsuzawa et al. 2004), suggesting a unique or novel receptor on the DNT-sensitive cells.

PMT is known as a powerful mitogen for fibroblasts (Rozengurt et al. 1990; Mullan and Lax 1996) and supports the anchorage-independent growth of Rat-1 fibroblasts (Higgins et al. 1992). It was also reported that PMT changes cell morphology that probably results from the aberrant formation of the actin cytoskeleton (Rutter and Luther 1984; Lacerda et al. 1996; Ohnishi et al. 1998). Like DNT, PMT also inhibits the differentiation of osteoblastic cells (Sterner-Kock et al. 1995; Mullan and Lax 1996). In addition, many reports have indicated that PMT induces the differentiation of osteoclasts (Gwaltney et al. 1997; Mullan and Lax 1998). However, it remains unclear whether PMT directly influences osteoclast lineage cells, an issue discussed in the next chapter. The receptor for PMT is not known. Although some reports suggested lipid components including gangliosides, sphingophospholipids, and phosphatidylcholine to be the receptor or co receptor for PMT (Pettit et al. 1993; Brothers et al. 2011), this has not been confirmed.

4.2 Effects of DNT and PMT on Osteogenesis-Related Cells and Tissues

Histological analyses of the nasal turbinate in AR revealed both bone resorption and the degeneration of osteoblasts and osteocytes (Rutter 1985). In experimental models, *B. bronchiseptica* or DNT induced pathological changes only in osteoblasts or osteocytes (Fetter et al. 1975; Silveira et al. 1982; Horiguchi et al. 1995a), while *P. multocida* or PMT stimulated osteoclastic activity and impaired osteoblastic bone formation (Foged et al. 1987; Kimman et al. 1987; Martineau-Doizé et al. 1993; Ghoshal and Niyo 1993; Mullan and Lax 1996). These phenomena were reproduced in experiments in vitro; Purified DNT affected the ability of osteoblastic MC3T3-E1 cells to differentiate from osteoprogenitors into osteocytes (Horiguchi et al. 1991). PMT inhibited the osteoblastic differentiation in osteosarcoma ROS 25/1 cells or chick bone-derived osteoblastic cells (Sterner-Kock et al. 1995; Mullan and Lax 1996). PMT also stimulated osteoclastic differentiation in various in vitro systems such as co cultures of osteoblasts and osteoclasts and bone marrow cells (Jutras and Martineau-Doizé 1996; Gwaltney et al. 1997; Mullan and Lax 1998; Hildebrand et al. 2011). Experimental evidence indicates that PMT acts directly on osteoblastic cells. However, the osteoclastic differentiation stimulated by PMT likely requires some population of cells other than osteoclasts.

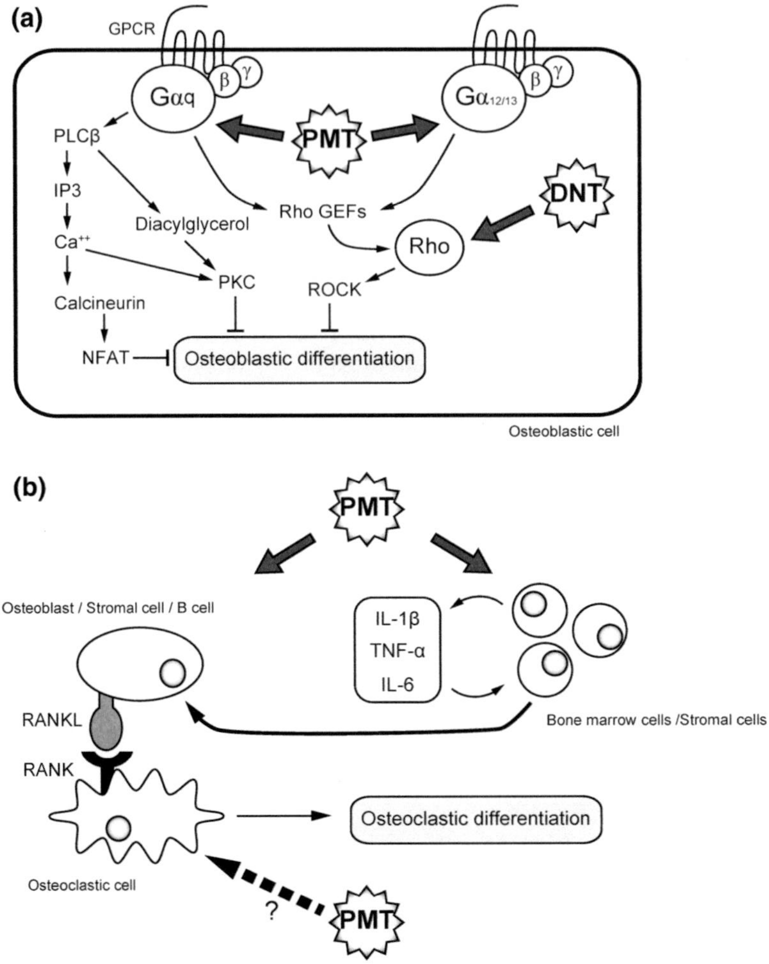

Fig. 5 The mechanisms by which PMT and DNT cause bone lesions. The bone lesions observed in the nasal turbinate and nasal septum where the bacteria colonize result from inhibition of osteoblastogenesis and stimulation of osteoclastogenesis. **a** Intracellular signaling pathways involved in osteoblastogenesis and affected by PMT and DNT. It is considered that ROCK, PKC, and NFAT eventually activated by PMT and DNT inhibit osteoblastogenesis although the mechanisms downstream of these signaling molecules still remain to be elucidated. **b** Proposed pathway of PMT-stimulated osteoclastogenesis. PMT stimulates some populations of bone marrow cells or stromal cells to secrete cytokines which upregulate RANKL expression in a paracrine or autocrine manner. RANKL induces osteoclastic lineage cells to differentiate via RANKL–RANK interaction. It is unknown whether PMT directly affects osteoclastic cells

5 Molecular Mechanisms Leading to Bone Lesions

Experimental evidence from histological and cell biological studies indicates turbinate atrophy to be attributable to impaired osteoblastic differentiation and bone formation, and accelerated osteoclastic differentiation and bone resorption in the osseous tissues. Both PMT and DNT inhibit osteoblastogenesis and, in addition, PMT stimulates osteoclastogenesis. As mentioned above, DNT activates the small GTPases of the Rho family while PMT activates α-subunits of G_i, G_q and $G_{12/13}$. We can therefore assume that downstream signaling pathways are involved in the turbinate atrophy. Intracellular signaling pathways influenced by DNT and PMT are summarized in Fig. 5. Reports indicate that some of the pathways regulate osteogenesis. The aberrant stimulation of these signaling pathways is considered to lead to disorders of bone formation and eventually the atrophy of turbinate bones.

5.1 Inhibition of Osteoblastogenesis by the Toxins

The inhibition of osteoblastic differentiation probably results from activation of Rho and its downstream effector ROCK (also called ROK or Rho-kinase) (Fig. 5a) (Horiguchi et al. 1991, 1995b; Sterner-Kock et al. 1995; Mullan and Lax 1996; Gwaltney et al. 1997; Harmey et al. 2004; Lax et al. 2004). Rho GTPase is directly modified and activated by DNT as mentioned above. PMT also activates Rho through G_q and/or $G_{12/13}$ heterotrimeric GTPases, probably via specific Rho GEFs including p63RhoGEF, p115RhoGEF, PDZ-RhoGEF, and LARG (Orth et al. 2005). C3 transferase, a Rho inhibitor, and the ROCK inhibitors HA-1077 and Y-27632 not only prevented the inhibitory effect of PMT on osteoblastic differentiation, but also stimulated osteogenesis in the absence of the toxin, indicating that the Rho-ROCK pathway negatively regulates the osteogenesis (Harmey et al. 2004). In addition, N-acetylcystein down-regulated Rho and stimulated osteoblastic differentiation, an effect which PMT suppressed (Jun et al. 2008). These results indicate the involvement of the Rho-ROCK pathway in osteoblastic differentiation. However, the underlying mechanism acting downstream of ROCK to regulate osteoblastic differentiation remains unknown.

In addition, PKC activated by G_q is implicated in the inhibitory effect of PMT on osteoblastic differentiation. It was reported that PMT stimulated PKC via an increase in intracellular DAG produced by PLCβ downstream of G_q (Staddon et al. 1990). Recently, several forms of PKC including PKCα were shown to negatively regulate osteoblastic differentiation (Ogata et al. 2007; Nakura et al. 2011) (Fig. 5a).

Furthermore, the nuclear factor of activated T cells (NFAT) might be also involved in the inhibition of osteoblastogenesis (Boss et al. 1996; Koga et al. 2005; Ogata et al. 2007; Choo et al. 2009; Jensen et al. 2010) (Fig. 5a). NFAT could be activated by PMT through G_q-mediated activation of PLCβ and, in turn, increases in intracellular Ca^{2+} and the activation of calcineurin (Fig. 5) (Aminova and Wilson 2007; Kamitani et al. 2010). Activated NFAT enhanced the transcriptional activity of osterix for *Col1a1*, which encodes type I collagen (Koga et al. 2005), but repressed the gene expression of osterix and osteocalcin, early and late markers of osteoblastic differentiation, respectively (Choo et al. 2009). Thus, NFAT could be a negative regulator for osteoblastic differentiation, although its role of NFAT in osteoblastogenesis is still controversial (Sitara and Aliprantis 2010).

5.2 Stimulation of Osteoclastogenesis by PMT

PMT stimulates osteoclastic bone resorption in vivo and in vitro (Elling and Pedersen 1985; Dominick and Rimler 1986, 1988; Kimman et al. 1987; Felix et al. 1992; Martineau-Doizé et al. 1993). Reports suggest PMT-induced differentiation of osteoclasts to be mediated by certain types of cells such as hemopoietic bone marrow cells (Jutras and Martineau-Doizé 1996; Gwaltney et al. 1997) and osteoblasts (Mullan and Lax 1998), and not to result from the direct action of the toxin. The receptor activator of NF-κB ligand (RANKL) is a key cytokine for osteoclastogenesis (Fig. 5b) (Takayanagi 2007; Findlay and Atkins 2011). It is known that the binding of RANKL, which is produced by osteoblasts and osteocytes, and other mesenchymal lineage cells, to its receptor RANK on myeloid osteoclast precursors promotes osteoclastic differentiation and bone resorbing activity. In osteoblast and osteoclast co cultures allowing cell–cell contact, PMT stimulated bone resorption (Mullan and Lax 1998). This phenomenon was not observed in co culture, in which each cell type was separated by porous membranes, implying the direct binding of RANKL on osteoblasts and RANK on osteoclast precursors to be required for PMT-induced osteoclastogenesis. More recently, it was demonstrated that B cells support PMT-induced differentiation of macrophages into mature osteoclasts (Hildebrand et al. 2011). In addition, PMT was found to stimulate the generation of B cells from bone marrow cells and the production of osteoclastogenic factors such as IL-1β, IL-6, TNF-α, and RANKL. IL-1β, IL-6, and TNF-α were reported to upregulate RANKL production in a paracrine and autocrine manner (Rogers and Eastell 2005). Taken together, one can assume that PMT stimulates mesenchymal cells, including B cells, to produce the osteoclastogenic cytokines, which in turn increase the expression of RANKL. RANKL binds to RANK on osteoclastic precursors and promotes osteoclastic differentiation (Fig. 5b). It is likely that osteoblasts, as well as B cells, take part in this cascade (Mullan and Lax 1998). It remains unknown which intracellular signaling pathways are affected by PMT to stimulate cytokine production. However, given that PMT targets heterotrimeric GTPases and the different types of targets are not known, a PMT-sensitive Gα subunit may, or should, be involved.

Interestingly, NFAT (NFATc1) in osteoclasts is known as a key positive regulator for osteoclastic differentiation. NFATc1 expression is induced by

RANKL–RANK interaction. The activation of NFAT is mediated by calcineurin, which is located downstream of a $G\alpha_q$-dependent signaling pathway (Boss et al. 1996). Therefore, one cannot rule out the possibility that cells of the osteoclastic lineage differentiate into osteoclasts in response to the direct actions of PMT (Fig. 5). However, it remains to be elucidated whether osteoclastic cells are sensitive to PMT.

6 Concluding Remarks

Several issues regarding PMT and DNT remain to be resolved. For example, the mechanisms by which *P. multocida* and *B. bronchiseptica* regulate the transcription of the toxin genes and secrete or release the toxins from the bacterial cells are unknown. The receptor molecules for these toxins have not been clearly identified either, which might be important to distinguish toxin-sensitive cells from resistant cells in host tissues. Nonetheless, a great deal of progress has been made in understanding the molecular actions of PMT and DNT, enabling speculation on how these toxins cause the characteristic pathological changes in nonprogressive AR and progressive AR. Furthermore, the fact that PMT and DNT, which activate specific small GTPases or heterotrimeric GTPases, cause bone lesions indicate these signaling pathways to be involved in the osteogenesis. Thus, these toxins could be not only protective antigens against the disease, but also useful probes for exploring the relationship between osteogenesis and these signaling pathways.

References

Aminova LR, Wilson BA (2007) Calcineurin-independent inhibition of 3T3-L1 adipogenesis by *Pasteurella multocida* toxin: suppression of Notch1, stabilization of beta-catenin and pre-adipocyte factor 1. Cell Microbiol 9:2485–2496

Aminova LR, Luo S, Bannai Y, Ho M, Wilson BA (2008) The C3 domain of *Pasteurella multocida* toxin is the minimal domain responsible for activation of Gq-dependent calcium and mitogenic signaling. Protein Sci 17:945–949

Baldwin MR, Lakey JH, Lax AJ (2004) Identification and characterization of the *Pasteurella multocida* toxin translocation domain. Mol Microbiol 54:239–250

Boss V, Talpade DJ, Murphy TJ (1996) Induction of NFAT-mediated transcription by Gq-coupled receptors in lymphoid and non-lymphoid cells. J Biol Chem 271:10429–10432

Brothers MC, Ho M, Maharjan R, Clemons NC, Bannai Y, Waites MA, Faulkner MJ, Kuhlenschmidt TB, Kuhlenschmidt MS, Blanke SR, Rienstra CM, Wilson BA (2011) Membrane interaction of *Pasteurella multocida* toxin involves sphingomyelin. FEBS J 278: 4633–4648

Busch C, Orth J, Djouder N, Aktories K (2001) Biological activity of a C-terminal fragment of *Pasteurella multocida* toxin. Infect Immun 69:3628–3634

Buys W, Smith H, Kamps A, Kamp E, Smits M (1990) Sequence of the dermonecrotic toxin of *Pasteurella multocida* ssp. *multocida*. Nucleic Acids Res 18:2815

Caprioli A, Falbo V, Roda LG, Ruggeri FM, Zona C (1983) Partial purification and characterization of an *Escherichia coli* toxic factor that induces morphological cell alterations. Infect Immun 39:1300–1306

Choo M-K, Yeo H, Zayzafoon M (2009) NFATc1 mediates HDAC-dependent transcriptional repression of osteocalcin expression during osteoblast differentiation. Bone 45:579–589

de Jong M (1983) Atrophic rhinitis caused by intranasal or intramuscular administration of broth-culture and broth-culture filtrates containing AR toxin of *Pasteurella multocida*. In: Pedersen KB, Nielsen NC (eds) Atrophic rhinitis in pigs. Office for official publications of the EEC, luxembourg, pp 136–146

de Jong MF, Nielsen JP (1990) Definition of progressive atrophic rhinitis. Vet Rec 126:93

Dominick MA, Rimler RB (1986) Turbinate atrophy in gnotobiotic pigs intranasally inoculated with protein toxin isolated from type D *Pasteurella multocida*. Am J Vet Res 47:1532–1536

Dominick MA, Rimler RB (1988) Turbinate osteoporosis in pigs following intranasal inoculation of purified *Pasteurella* toxin: histomorphometric and ultrastructural studies. Vet Pathol 25:17–27

Elling F, Pedersen KB (1985) The pathogenesis of persistent turbinate atrophy induced by toxigenic *Pasteurella multocida* in pigs. Vet Pathol 22:469–474

Felix R, Fleisch H, Frandsen PL (1992) Effect of *Pasteurella multocida* toxin on bone resorption in vitro. Infect Immun 60:4984–4988

Fetter AW, Switzer WP, Capen CC (1975) Electron microscopic evaluation of bone cells in pigs with experimentally induced *Bordetella* rhinitis (turbinate osteoporosis). Am J Vet Res 36:15–22

Findlay DM, Atkins GJ (2011) Relationship between serum RANKL and RANKL in bone. Osteoporos Int 22:2597–2602

Foged N, Pedersen K, Elling F (1987) Characterization and biological effects of the *Pasteurella multocida* toxin. FEMS Microbiol Lett 43:45–51

Fukui A, Horiguchi Y (2004) *Bordetella* dermonecrotic toxin exerting toxicity through activation of the small GTPase Rho. J Biochem 136:415–419

Fukui-Miyazaki A, Ohnishi S, Kamitani S, Abe H, Horiguchi Y (2011) *Bordetella* dermonecrotic toxin binds to target cells via the N-terminal 30 amino acids. Microbiol Immunol 55:154–159

Ghoshal NG, Niyo Y (1993) Histomorphologic features of the nasal cavity of pigs exposed to *Pasteurella multocida* type-D dermonecrotic toxin. Am J Vet Res 54:738–742

Gwaltney SM, Galvin RJS, Register KB, Rimler RB, Ackermann MR (1997) Effects of *Pasteurella multocida* toxin on porcine bone marrow cell differentiation into osteoclasts and osteoblasts. Vet Pathol 34:421–430

Hanada M, Shimoda K, Tomita S, Nakase Y, Nishiyama Y (1979) Production of lesions similar to naturally occurring swine atrophic rhinitis by cell-free sonicated extract of *Bordetella bronchiseptica*. Nippon Juigaku Zasshi 41:1–8

Harmey D, Stenbeck G, Nobes CD, Lax AJ, Grigoriadis AE (2004) Regulation of osteoblast differentiation by *Pasteurella multocida* toxin (PMT): a role for Rho GTPase in bone formation. J Bone Miner Res 19:661–670

Harris DL, Switzer WP (1968) Turbinate atrophy in young pigs exposed to *Bordetella bronchiseptica*, *Pasteurella multocida*, and combined inoculum. Am J Vet Res 29:777–785

Higgins TE, Murphy AC, Staddon JM, Lax AJ, Rozengurt E (1992) *Pasteurella multocida* toxin is a potent inducer of anchorage-independent cell growth. Proc Natl Acad Sci U S A 89:4240–4244

Hildebrand D, Heeg K, Kubatzky KF (2011) *Pasteurella multocida* toxin-stimulated osteoclast differentiation is B cell dependent. Infect Immun 79:220–228

Hoffmann C, Schmidt G (2004) CNF and DNT. Rev Physiol Biochem Pharmacol 152:49–63

Horiguchi Y, Nakai T, Kume K (1989) Purification and characterization of *Bordetella bronchiseptica* dermonecrotic toxin. Microb Pathog 6:361–368

Horiguchi Y, Nakai T, Kume K (1991) Effects of *Bordetella bronchiseptica* dermonecrotic toxin on the structure and function of osteoblastic clone MC3T3-e1 cells. Infect Immun 59:1112–1116

Horiguchi Y, Sugimoto N, Matsuda M (1993) Stimulation of DNA synthesis in osteoblast-like MC3T3-E1 cells by *Bordetella bronchiseptica* dermonecrotic toxin. Infect Immun 61:3611–3615

Horiguchi Y, Sugimoto N, Matsuda M (1994) *Bordetella bronchiseptica* dermonecrotizing toxin stimulates protein synthesis in an osteoblastic clone, MC3T3-E1 cells. FEMS Microbiol Lett 120:19–22

Horiguchi Y, Okada T, Sugimoto N, Morikawa Y, Katahira J, Matsuda M (1995a) Effects of *Bordetella bronchiseptica* dermonecrotizing toxin on bone formation in calvaria of neonatal rats. FEMS Immunol Med Microbiol 12:29–32

Horiguchi Y, Senda T, Sugimoto N, Katahira J, Matsuda M (1995b) *Bordetella bronchiseptica* dermonecrotizing toxin stimulates assembly of actin stress fibers and focal adhesions by modifying the small GTP-binding protein rho. J Cell Sci 108(Pt 10):3243–3251

Horiguchi Y, Inoue N, Masuda M, Kashimoto T, Katahira J, Sugimoto N, Matsuda M (1997) *Bordetella bronchiseptica* dermonecrotizing toxin induces reorganization of actin stress fibers through deamidation of Gln-63 of the GTP-binding protein Rho. Proc Natl Acad Sci U S A 94:11623–11626

Jensen ED, Gopalakrishnan R, Westendorf JJ (2010) Regulation of gene expression in osteoblasts. BioFactors 36(1):25–32

Jun JH, Lee S-H, Kwak HB, Lee ZH, Seo S-B, Woo KM, Ryoo H-M, Kim G-S, Baek J-H (2008) N-acetylcysteine stimulates osteoblastic differentiation of mouse calvarial cells. J Cell Biochem 103:1246–1255

Jutras I, Martineau-Doizé B (1996) Stimulation of osteoclast-like cell formation by *Pasteurella multocida* toxin from hemopoietic progenitor cells in mouse bone marrow cultures. Can J Vet Res 60:34–39

Kamitani S, Kitadokoro K, Miyazawa M, Toshima H, Fukui A, Abe H, Miyake M, Horiguchi Y (2010) Characterization of the membrane-targeting C1 Domain in *Pasteurella multocida* toxin. J Biol Chem 285:25467–25475

Kamitani S, Ao S, Toshima H, Tachibana T, Hashimoto M, Kitadokoro K, Fukui-Miyazaki A, Abe H, Horiguchi Y (2011) Enzymatic actions of *Pasteurella multocida* toxin detected by monoclonal antibodies recognizing the deamidated α subunit of the heterotrimeric GTPase Gq. FEBS J 278:2702–2712

Kashimoto T, Katahira J, Cornejo WR, Masuda M, Fukuoh A, Matsuzawa T, Ohnishi T, Horiguchi Y (1999) Identification of functional domains of *Bordetella* dermonecrotizing toxin. Infect Immun 67:3727–3732

Kimman TG, Löwik CW, van de Wee-Pals LJ, Thesingh CW, Defize P, Kamp EM, Bijvoet OL (1987) Stimulation of bone resorption by inflamed nasal mucosa, dermonecrotic toxin-containing conditioned medium from *Pasteurella multocida*, and purified dermonecrotic toxin from *P. multocida*. Infect Immun 55:2110–2116

Kitadokoro K, Kamitani S, Miyazawa M, Hanajima-Ozawa M, Fukui A, Miyake M, Horiguchi Y (2007) Crystal structures reveal a thiol protease-like catalytic triad in the C-terminal region of *Pasteurella multocida* toxin. Proc Natl Acad Sci U S A 104:5139–5144

Knust Z, Schmidt G (2010) Cytotoxic necrotizing factors (CNFs)—a growing toxin family. Toxins 2:116–127

Koga T, Matsui Y, Asagiri M, Kodama T, de Crombrugghe B, Nakashima K, Takayanagi H (2005) NFAT and osterix cooperatively regulate bone formation. Nat Med 11:880–885

Lacerda HM, Lax AJ, Rozengurt E (1996) *Pasteurella multocida* toxin, a potent intracellularly acting mitogen, induces p125FAK and paxillin tyrosine phosphorylation, actin stress fiber formation, and focal contact assembly in Swiss 3T3 cells. J Biol Chem 271:439–445

Lax AJ, Chanter N, Pullinger GD, Higgins T, Staddon JM, Rozengurt E (1990) Sequence analysis of the potent mitogenic toxin of *Pasteurella multocida*. FEBS Lett 277:59–64

Lax AJ, Pullinger GD, Baldwin MR, Harmey D, Grigoriadis AE, Lakey JH (2004) The *Pasteurella multocida* toxin interacts with signalling pathways to perturb cell growth and differentiation. Int J Med Microbiol 293:505–512

Lemichez E, Flatau G, Bruzzone M, Boquet P, Gauthier M (1997) Molecular localization of the *Escherichia coli* cytotoxic necrotizing factor CNF1 cell-binding and catalytic domains. Mol Microbiol 24:1061–1070

Martineau-Doizé B, Caya I, Gagné S, Jutras I, Dumas G (1993) Effects of *Pasteurella multocida* toxin on the osteoclast population of the rat. J Comp Pathol 108:81–91

Masuda M, Betancourt L, Matsuzawa T, Kashimoto T, Takao T, Shimonishi Y, Horiguchi Y (2000) Activation of rho through a cross-link with polyamines catalyzed by *Bordetella* dermonecrotizing toxin. EMBO J 19:521–530

Matsuzawa T, Kashimoto T, Katahira J, Horiguchi Y (2002) Identification of a receptor-binding domain of *Bordetella* dermonecrotic toxin. Infect Immun 70:3427–3432

Matsuzawa T, Fukui A, Kashimoto T, Nagao K, Oka K, Miyake M, Horiguchi Y (2004) *Bordetella* dermonecrotic toxin undergoes proteolytic processing to be translocated from a dynamin-related endosome into the cytoplasm in an acidification-independent manner. J Biol Chem 279:2866–2872

Mullan PB, Lax AJ (1996) *Pasteurella multocida* toxin is a mitogen for bone cells in primary culture. Infect Immun 64:959–965

Mullan PB, Lax AJ (1998) *Pasteurella multocida* toxin stimulates bone resorption by osteoclasts via interaction with osteoblasts. Calcif Tissue Int 63:340–345

Murphy AC, Rozengurt E (1992) *Pasteurella multocida* toxin selectively facilitates phosphatidylinositol 4,5-bisphosphate hydrolysis by bombesin, vasopressin, and endothelin. Requirement for a functional G protein. J Biol Chem 267:25296–25303

Nagai S, Someno S, Yagihashi T (1994) Differentiation of toxigenic from nontoxigenic isolates of *Pasteurella multocida* by PCR. J Clin Microbiol 32:1004–1010

Nakai T, Sawata A, Tsuji M, Kume K (1984a) Characterization of dermonecrotic toxin produced by serotype D strains of *Pasteurella multocida*. Am J Vet Res 45:2410–2413

Nakai T, Sawata A, Tsuji M, Samejima Y, Kume K (1984b) Purification of dermonecrotic toxin from a sonic extract of *Pasteurella multocida* SP-72 serotype D. Infect Immun 46:429–434

Nakura A, Higuchi C, Yoshida K, Yoshikawa H (2011) PKCα suppresses osteoblastic differentiation. Bone 48:476–484

Ogata N, Kawaguchi H, Chung U-I, Roth SI, Segre GV (2007) Continuous activation of G alpha q in osteoblasts results in osteopenia through impaired osteoblast differentiation. J Biol Chem 282:35757–35764

Ohnishi T, Horiguchi Y, Masuda M, Sugimoto N, Matsuda M (1998) *Pasteurella multocida* toxin and *Bordetella bronchiseptica* dermonecrotizing toxin elicit similar effects on cultured cells by different mechanisms. J Vet Med Sci 60:301–305

Orth JHC, Blöcker D, Aktories K (2003) His1205 and His1223 are essential for the activity of the mitogenic *Pasteurella multocida* toxin. Biochemistry 42:4971–4977

Orth JHC, Lang S, Taniguchi M, Aktories K (2005) *Pasteurella multocida* toxin-induced activation of RhoA is mediated via two families of Gα proteins, Gαq and Gα12/13. J Biol Chem 280:36701–36707

Orth JHC, Fester I, Preuss I, Agnoletto L, Wilson BA, Aktories K (2008) Activation of Galpha (i) and subsequent uncoupling of receptor-Galpha(i) signaling by *Pasteurella multocida* toxin. J Biol Chem 283:23288–23294

Orth JHC, Preuss I, Fester I, Schlosser A, Wilson BA, Aktories K (2009) *Pasteurella multocida* toxin activation of heterotrimeric G proteins by deamidation. Proc Natl Acad Sci U S A 106:7179–7184

Oswald E, Sugai M, Labigne A, Wu HC, Fiorentini C, Boquet P, O'Brien AD (1994) Cytotoxic necrotizing factor type 2 produced by virulent *Escherichia coli* modifies the small GTP-binding proteins Rho involved in assembly of actin stress fibers. Proc Natl Acad Sci U S A 91:3814–3818

Pedersen KB, Barfod K (1981) The aetiological significance of *Bordetella bronchiseptica* and *Pasteurella multocida* in atrophic rhinitis of swine. Nord Vet Med 33:513–522

Pedersen KB, Elling F (1984) The pathogenesis of atrophic rhinitis in pigs induced by toxigenic *Pasteurella multocida*. J Comp Pathol 94:203–214

Petersen SK (1990) The complete nucleotide sequence of the *Pasteurella multocida* toxin gene and evidence for a transcriptional repressor, TxaR. Mol Microbiol 4:821–830

Pettit RK, Ackermann MR, Rimler RB (1993) Receptor-mediated binding of *Pasteurella multocida* dermonecrotic toxin to canine osteosarcoma and monkey kidney (vero) cells. Lab Invest 69:94–100

Pullinger GD, Sowdhamini R, Lax AJ (2001) Localization of functional domains of the mitogenic toxin of *Pasteurella multocida*. Infect Immun 69:7839–7850

Pullinger GD, Bevir T, Lax AJ (2004) The *Pasteurella multocida* toxin is encoded within a lysogenic bacteriophage. Mol Microbiol 51:255–269

Rogers A, Eastell R (2005) Circulating osteoprotegerin and receptor activator for nuclear factor kappaB ligand: clinical utility in metabolic bone disease assessment. J Clin Endocrinol Metab 90:6323–6331

Rozengurt E, Higgins T, Chanter N, Lax AJ, Staddon JM (1990) *Pasteurella multocida* toxin: potent mitogen for cultured fibroblasts. Proc Natl Acad Sci U S A 87:123–127

Rutter JM (1985) Atrophic rhinitis in swine. Adv Vet Sci Comp Med 29:239–279

Rutter JM, Luther PD (1984) Cell culture assay for toxigenic *Pasteurella multocida* from atrophic rhinitis of pigs. Vet Rec 114:393–396

Rutter J, Mackenzie A (1984) Pathogenesis of atrophic rhinitis in pigs: a new perspective. Vet Rec 114:89

Silveira D, Edington N, Smith IM (1982) Ultrastructural changes in the nasal turbinate bones of pigs in early infection with *Bordetella bronchiseptica*. Res Vet Sci 33:37–42

Sitara D, Aliprantis AO (2010) Transcriptional regulation of bone and joint remodeling by NFAT. Immunol Rev 233:286–300

Staddon JM, Chanter N, Lax AJ, Higgins TE, Rozengurt E (1990) *Pasteurella multocida* toxin, a potent mitogen, stimulates protein kinase C-dependent and -independent protein phosphorylation in Swiss 3T3 cells. J Biol Chem 265:11841–11848

Sterner-Kock A, Lanske B, Uberschar S, Atkinson MJ (1995) Effects of the *Pasteurella multocida* toxin on osteoblastic cells in vitro. Vet Pathol 32:274–279

Takayanagi H (2007) The role of NFAT in osteoclast formation. Ann NY Acad Sci 1116:227–237

Walker KE, Weiss AA (1994) Characterization of the dermonecrotic toxin in members of the genus *Bordetella*. Infect Immun 62:3817–3828

Ward PN, Miles AJ, Sumner IG, Thomas LH, Lax AJ (1998) Activity of the mitogenic *Pasteurella multocida* toxin requires an essential C-terminal residue. Infect Immun 66:5636–5642

Zywietz A, Gohla A, Schmelz M, Schultz G, Offermanns S (2001) Pleiotropic effects of *Pasteurella multocida* toxin are mediated by Gq-dependent and -independent mechanisms. Involvement of Gq but not G11. J Biol Chem 276:3840–3845

The *Pasteurella multocida* Toxin: A New Paradigm for the Link Between Bacterial Infection and Cancer

Alistair Lax

Abstract The concept that bacterial infection could cause cancer has only recently become accepted because of the strong epidemiological and molecular evidence for a major carcinogenic role played by *Helicobacter pylori*. However, information on other potential bacterial carcinogens is very limited and thereby unconvincing. A different approach is to assess bacteria for potentially pro-carcinogenic properties. The *Pasteurella multocida* toxin (PMT) has many properties that mark it out as a potential carcinogen. PMT is a highly potent mitogen and has been demonstrated to block apoptosis. PMT modifies and activates members of three of the four families of heterotrimeric G-proteins, all of which have potential roles in carcinogenesis. Many signalling components downstream of these G-proteins are known proto-oncogenes and have been shown to be activated by PMT. These include, amongst others, the Rho GTPase, focal adhesion kinase, cyclooxygenase-2, β-catenin signalling and calcium signalling. PMT action potentially influences many of the acquired Hanahan/ Weinberg capabilities necessary for oncogenic transformation. Although there is little evidence that PMT might have a role in human cancer, it serves as an important and novel paradigm for a bacterial link to cancer.

Contents

A. Lax (✉)
Department of Microbiology, King's College London Dental Institute, London, UK
e-mail: alistair.lax@kcl.ac.uk

Current Topics in Microbiology and Immunology (2012) 361: 131–144
DOI: 10.1007/82_2012_236

Published Online: 15 June 2012

The concept that bacteria could play a role in the causation of cancer has evolved slowly, but has recently become more widely accepted for a limited number of bacteria. The initial suggestion that bacterial infection could predispose to cancer came shortly after the birth of medical microbiology in the 1870s that was led by Robert Koch, Louis Pasteur and their associates. Although others pre-date him, the Scottish pathologist William Russell is often noted as the first to suggest that bacteria at the site of a tumour contribute to carcinogenesis. It is, therefore, rather bizarre that his oft quoted paper reports the presence of a fungal pathogen in many cancers (Russell 1890). However, there was little main stream scientific support for the concept that these organisms were in any way causative until the demonstration in the early 1990s that *Helicobacter pylori* infection led to an increased risk of gastric adenocarcinoma (Nomura et al. 1991; Parsonnet et al. 1991). *Helicobacter* was also shown to cause gastric mucosa associated lymphoid tissue (MALT) lymphoma (Wotherspoon et al. 1991).The subsequent 20 years of work on *Helicobacter* have identified several bacterial virulence determinants that activate multiple signalling pathways (Franco et al. 2005; Tegtmeyer et al. 2011). These advances, along with substantial epidemiological evidence and experimental animal studies, provide convincing support for a causative link between *Helicobacter* and cancer, where it is recognised that both inflammatory and non-inflammatory mechanisms participate.

A few other bacteria have a proven epidemiological link to human cancer. Notably, long term carriage of *Salmonella typhi* is known to lead to hepatobiliary carcinoma (Welton et al. 1979; Caygill et al. 1995), although the molecular mechanisms remain unknown. Several bacteria are strongly implicated in specific MALT lymphomas. For example, *Chlamydia* infection is linked to ocular adnexal lymphomas (Ferreri et al. 2009). Evidence for the involvement of other bacteria in malignant disease is suggestive, but not proven. What is frequently missing is molecular evidence of a carcinogenic phenotype. There has also been a more general hypothesis suggesting that infections contribute significantly to carcinogenesis, with inflammation being a major contributory factor (Kuper et al. 2000).

A new approach to this subject in recent years has been to examine bacteria, in particular their toxins, whose biochemical and cellular effects suggest that they display a pro-carcinogenic signature (Lax and Thomas 2002; Lax 2005; Oswald et al. 2005). Discussion of cancer and its causation relies on two related paradigms. First is the notion that carcinogenic agents can act to either initiate or promote carcinogenesis, leading to a gradual sequential accumulation of deregulatory mutations. The second concept, that was first precisely framed by Hanahan and Weinberg (2000) and recently revised (Hanahan and Weinberg 2011), is that such mutations lead to a limited and defined set of acquired capabilities that a cell must gain to become transformed. It is against the criteria defined by these two overarching concepts that any potentially carcinogenic agent must be tested.

Several bacteria and their toxins display characteristics that fulfil these pro-carcinogenic criteria. The *Pasteurella multocida* toxin (PMT) was one of the first toxins whose effects on cells suggested strongly that it could fall into this category. It is the purpose of this chapter to show to what extent the known biochemical and cellular effects of PMT mark it out as having a carcinogenic signature. In addition, it is pertinent to consider whether it is likely that PMT could have a role in the causation of human cancer.

1 *Pasteurella multocida* Toxin: A Modulator of G-Protein Linked Intracellular Signalling

PMT was identified through its role in causing porcine atrophic rhinitis (Pederson and Barfod 1981), a disease specifically associated with loss of nasal bone in developing piglets (Switzer and Farrington 1975). Early experiments with cultured cells suggested that PMT produced a cytotoxic phenotype (Rutter and Luther 1984). The toxin gene was then cloned and expressed by several groups, and this advance enabled more detailed analysis of its function. Analysis of the effect of recombinant PMT on cultured Swiss 3T3 cells led to the startling and unexpected conclusion that PMT could operate as a highly potent mitogen that was able to stimulate quiescent cells to re-enter the cell cycle, traverse S phase and undergo mitosis (Rozengurt et al. 1990). Our subsequent early work demonstrated that several signalling pathways were activated, notably inositol trisphosphate signalling, calcium mobilisation and protein kinase C activation, along with pathways linked to cytoskeletal rearrangements and Rho GTPase activation (Staddon et al. 1990, 1991; Lacerda et al. 1996).

These first observations on signalling pathways affected by PMT action suggested that the toxin led to activation of phospholipase C. Experiments using GTP analogues strongly supported the concept that phospholipase C was stimulated via activation of the heterotrimeric G-protein G_q (Murphy and Rozengurt 1992; Wilson et al. 1997). As detailed elsewhere in this book, it was later demonstrated using G-protein deficient mouse embryonic fibroblasts that the G_{12} G-protein family was also activated by PMT (Zywietz et al. 2001; Orth et al. 2005), and more recently that members of the G_i family of G-proteins were the target for PMT action (Orth et al. 2008). Thus, the concept was established that PMT acts on three of the four subclasses of heterotrimeric G-proteins, G_s being the only exception. Nevertheless, not all members within each subclass are activated. For example, while G_q is activated by PMT, the closely related G_{11} is not (Zywietz et al. 2001). It is now known that the inability of PMT to activate G_{11} is not because the target amino acid in $G\alpha_q$ is missing in $G\alpha_{11}$ (Orth et al. 2009), but must reflect a difference in the context of the target amino acid (Orth et al. 2004). As detailed elsewhere in this book, the enzymatic action of PMT covalently modifies its target $G\alpha$ subunit to block the intrinsic GTPase of the G-protein and so lock it in a permanently active state. The protein sequence of PMT does not suggest that PMT has any other enzymatic activity. In support of this, it is noteworthy that a single amino acid substitution mutant at the PMT active site is entirely innocuous when injected into experimental animals at a thousand times the lethal dose of wild type toxin (Ward et al. 1998).

The heterotrimeric G-proteins are composed of α, β and γ subunits and are classified into four families (G_q, G_{12}, G_i and G_s) according to which α subunit is present. Further subdivision within each group leads to a total of over 20 different α subunits, but not all subtypes are expressed in every cell. There are 5 $G\beta$ subtypes and 12 $G\gamma$ subtypes, so that in theory there are over a 1,000 possible combinations (Oldham and Hamm 2008), but the different roles that G-protein subtypes might play has only recently begun to be investigated (Krumins and Gilman 2006; Foley et al. 2010). G-proteins are attached to the inner surface of the cytoplasmic membrane and interact with the G-protein coupled receptors (GPCRs) that span the membrane. In contrast to the apparently limited family of G-proteins (as defined by the $G\alpha$ subunit), there are several hundred GPCRs, which provide the cellular interface that interacts with numerous ligands, including many pharmaceutical drugs and drugs of abuse (Lagerström and Schiöth 2008). Activation by a GPCR causes loss of GDP which is bound to the $G\alpha$ subunit and its replacement with GTP. The conformational change induced by GTP binding leads to separation of the α subunit from the $\beta\gamma$ complex. The α subunit, and in some cases also the $\beta\gamma$ subunit, then stimulate downstream signalling by interaction with intermediary signalling proteins. The intrinsic GTPase activity of the α subunit then cleaves GTP to GDP and the G-protein subunits reassociate to return the G-protein to the resting state. Various proteins interact with the α subunit to stimulate or inhibit the GTPase. As well as being regulated by GTP/GDP exchange, G-protein function is also affected by reversible tyrosine phosphorylation, reversible lipid modification and cellular localisation (Oldham and Hamm 2008).

The four G-protein families were originally linked to a distinct set of signalling pathways (Neves et al. 2002). Members of the G_s and G_i families each regulate the concentration of cyclic AMP in the cell; G_s stimulates adenylate cyclase, while G_i downregulates this activation. G_q family proteins, of which G_q and G_{11} are the most ubiquitously expressed variants, activate phospholipase $C\beta$ ($PLC\beta$) (Lee et al. 1992). G_{12} family proteins (G_{12} and G_{13}) stimulate the Rho protein by interaction with specific Rho guanine nucleotide exchange factors (RhoGEFs) (Tanabe et al. 2004). There is, however, increasing awareness of the overlap between the G-protein stimulated pathways, and new interconnections are still being identified. For example, G_q activates Rho (Chikumi et al. 2002), G-proteins of the G_{12} family interact directly with E-cadherin to influence β-catenin activity (Meigs et al. 2001) and G_i has been shown to regulate c-Met tyrosine kinase signalling (Zhao et al. 2007). Moreover, it is now realised that G-proteins have other interactions outside these pathways and at other locations (Hewavitharana and Wedegaertner 2012).

Thus the G-proteins are key players in eukaryotic signal transduction, and involved in many critical cellular functions, including the regulation of cell division and growth, apoptosis and differentiation and metabolic regulation. It is therefore not surprising that a toxin that can activate members of three of the four families of G-proteins has such an affect on cellular function.

2 Evidence for a Pro-Carcinogenic Role for PMT

It did not escape our notice in 1990 when we identified the mitogenic properties of PMT (Rozengurt et al. 1990) that these properties were potentially pro-carcinogenic. Indeed, earlier work by Rutter and Mackenzie (1984) had demonstrated that injection of experimental animals with purified PMT led to a variety of histopathological changes, including a notable hyperplasia in bladder and ureter transitional epithelia. These latter effects were reproduced with recombinant PMT (Lax and Chanter 1990), and we later showed that experimental infection with toxigenic *P. multocida* also led to hyperplasia in these tissues (Hoskins et al. 1997). There is no suggestion that the action of PMT directly leads to initiating mutations in either cultured cells or experimental animals, but there is significant evidence that marks it out as a potential tumour promoter. The G-proteins and their downstream signalling pathways are very strongly implicated in the carcinogenic process.

It is not surprising that all four G-protein families have transforming ability (Radhika and Dhanasekaran 2001). Activating mutations in either G_{12} or G_{13} have been shown to be transforming (Voyno-Yasenetskaya et al. 1994) and these G-proteins activate at least three oncogenic pathways. Rho signalling is activated via specific RhoGEFs (Tanabe et al. 2004), while direct interaction with E-cadherin interferes with adhesion (Meigs et al. 2002). As well as its role in adhesion, E-cadherin binds to the pro-mitogenic factor β-catenin, which is a key component of the canonical Wnt pathway. Inactivation of E-cadherin leads to release of β-catenin and its subsequent nuclear translocation. Activation of G_{12} proteins stimulates β-catenin release via the G-protein/E-cadherin interaction (Meigs et al. 2001). Aberrant E–cadherin expression is linked to several cancers including breast, prostate and oral carcinogenesis (Da Silva et al. 2008; Saha et al. 2008; Shah et al. 2009. Elevated cytoplasmic and nuclear β-catenin levels are implicated in the production of human colonic polyps which frequently become cancerous (Sparks et al. 1998). $G\alpha_i$ mutations have been identified in human tumours (Lyons et al. 1990). Similarly members of the G_q family have also been linked to transformation (Kalinec et al. 1992). G_q and G_{13} activation each lead to c-*jun* activation, which is critical for transformation via MAP kinase cascades (Marinissen et al. 2003). Moreover, G_q can downregulate β-catenin signalling by stimulating its calpain-mediated degradation via Ca^{2+}-mediated signalling (Li and Iyengar 2002). Mutations in $G\alpha_q$ have recently been found in a rare human cancer (Van Raamsdonk et al. 2009); indeed the pro-oncogenic mutation in G_q is the same as that produced by the action of PMT (Orth et al. 2009).

Each family of G-proteins has been shown to affect apoptosis, but the picture is confusing and there are examples for each G-protein of both pro- and anti-apoptotic effects (Yanamadala et al. 2009). The cellular choice between these opposing mechanisms may depend on cell type, the duration and strength of signal, downstream feedback mechanisms and whether the test cell is transformed or not. Most studies with G_{12} have demonstrated that its activation promotes apoptosis, as is also the case for the G_q family. A larger number of publications with G_i have found equal evidence for G_i either blocking or promoting apoptosis.

As might be predicted, activation of signalling components downstream of these G-proteins is also oncogenic. G_q activation stimulates phospholipase C leading to the generation of diacylglycerol which in turn activates protein kinase C (Stabel and Parker 1991). The protein kinase C family are serine/threonine kinases with multiple targets, and many members of this family are activated by the tumour promoting phorbol esters. Protein kinase Cε has multiple cellular effects, strongly suggesting it has a role in invasion and metastasis, and has been associated with several cancers (Basu and Sivaprasad 2007). The other product of phospholipase C activation is inositol trisphosphate, whose main action is to bind to receptors on intracellular calcium stores to modulate the release of Ca^{2+}. The elevation of intracellular Ca^{2+} has multiple effects on the proliferation and survival of cancer cells (Roderick and Cook 2008). PMT is known to stimulate protein kinase C activation (Rozengurt et al. 1990; Staddon et al. 1990), and to increase inositol 1,4,5-trisphosphate to stimulate Ca^{2+} release (Staddon et al. 1991).

The Rho family of small GTPases are activated by G_{12}, and to a lesser extent G_q, via specific Rho GEFs (guanine nucleotide exchange factors) that facilitate removal of GDP from Rho proteins and subsequent binding of GTP. The Rho protein family comprises three main members (A, B and C) that act as major cellular nodes for signalling pathways, with roles in cell proliferation, apoptosis and migration (Etienne-Manneville and Hall 2002). These effects translate into potential effects on several key characteristics of cancer (Titus et al. 2005; Narumiya et al. 2009). To date, activating mutations in the Rho proteins have not been found in human cancers, although up-regulation of Rho expression is associated with several cancers, with increased RhoC activation being strongly associated with metastasis (Clark et al. 2000; Kleer et al. 2002). Rho signals through several downstream pathways, including the Rho kinase ROCK, and this in turn activates other components including the focal adhesion kinase (FAK) and Cyclooxygenase-2 (Cox-2). FAK is involved in many cellular processes linked to cancer, including angiogenesis, migration, adhesion and growth, and its over expression and gene amplification have been described in many cancers (Golubovskaya 2010). Similarly, Cox-2 expression is widely accepted as having a major role in cancer, both in pre-cancer, where prophylactic inhibition of the Cox-2 enzyme is protective, and in cancer progression where Cox-2 inhibitors can delay the carcinogenic process (Greenhough et al. 2009). PMT is known to stimulate Rho-linked pathways through G-protein activation (Rozengurt et al. 1990: Lacerda et al. 1996). This effect leads not only to striking actin fibre rearrangements, but also via the activity of the Rho kinase to activation of the FAK which has links to human cancer (Agochiya et al. 1999). FAK in turn increases its interaction with the proto-oncogene Src (Thomas et al. 2001). It has been shown that PMT leads to increased expression of Cox-2 through activation of STAT transcription factors (Orth et al. 2007).

The main target of both G_s and G_i is the enzyme adenylate cyclase, whose activation leads to a rise in the intracellular messenger adenosine 3'5'cyclic monophosphate (cyclic AMP), and this in turn binds to regulatory subunits of protein kinase A (a serine/threonine kinase), leading to its activation. Although G_i acts to downregulate adenylate cyclase it has been linked to some cancers and the protein kinase A pathway has stimulatory or inhibitory effects on growth that are

Table 1 Signalling components affected by PMT that are involved in cancer

G-proteins	G_q, G_i, G_{12}
Intermediary signalling proteins	Protein kinase C Inositol/Ca^{2+} Rho Focal adhesion kinase Src COX-2 Protein kinase A β-catenin

cell-type specific. Protein kinase A is being considered as a marker and/or therapeutic target in cancer (Naviglio et al. 2009). In a similar manner, this pathway has both pro- and anti-apoptotic effects (Insel et al. 2012). Although it was noted several years ago that PMT affected cyclic AMP concentration in cells (Mullan and Lax 1996), this was not understood as a direct effect on G_i until much later (Orth et al. 2008). Given the variation in cellular response to the G_i/cyclic AMP/protein kinase A pathway, it is likely that the growth promoting or inhibiting effect of PMT will be cell-type specific (Table 1).

The canonical Wnt signalling pathway is linked to many cellular processes. Activation of this pathway leads to stabilisation of β-catenin, which enters the nucleus to drive proliferative gene expression (MacDonald et al. 2009). Dysregulation of the Wnt/β-catenin signalling pathway is strongly correlated with various cancers (White et al. 2012). PMT has been shown to stabilise β-catenin (Aminova and Wilson 2007). It is noteworthy that this key pathway is also stimulated by *H. pylori* (Franco et al. 2005) and by the *B. fragilis* toxin (Wu et al. 2003).

At the cellular level, several effects of PMT mark it out as having a carcinogenic signature. The toxin is a highly potent mitogen for several cell types (Rozengurt et al. 1990), and has also been shown to block apoptosis (Sabri et al. 2002; Preuß et al. 2010). It is also able to strongly promote anchorage independent growth in soft agar (Higgins et al. 1992), a property that was linked to activation of the Pim-1/SOCS-1 pathway (Hildebrand et al. 2010). The toxin synergises with bombesin (Murphy and Rozengurt 1992), a neuropeptide known to have a link with small cell carcinoma of the lung (Woll and Rozengurt 1988). In addition, PMT inhibits differentiation and maturation of osteoblasts to form bone nodules (Harmey et al. 2004) and adipogenesis (Aminova and Wilson 2007).

3 The Potential of PMT to Influence the Hallmarks of Cancer

The original concept that cancer exhibited a defined set of hallmarks or capabilities (Hanahan and Weinberg 2000) has now been revised and expanded to include further capabilities (Hanahan and Weinberg 2011), suggesting that a cell has to acquire 8 such traits in order to transform into a fully dysregulated cancer cell.

Table 2 The Hallmarks of Cancer capabilities. The revised and updated set of acquire capabilities suggested by Hanahan and Weinberg (2011) to be required for cancer to develop

Revised	Self sufficiency in growth signals
	Insensitivity to antigrowth signals
	Evading apoptosis
	Limitless replicative potential
	Sustained angiogenesis
	Tissue invasion and metastasis
Updated	Deregulating cellular energetics
	Avoiding immune destruction

This theory is predicated on the concept that each of these capabilities becomes hard-wired into the cell by mutation. However, it is not suggested that PMT, or indeed most bacterial toxins with pro-carcinogenic potential, directly facilitates mutation, but rather that some of its effects mimic those of a pro-carcinogenic mutation. While the toxin stimulus is applied, the cell will behave as if it has acquired the capability and will be indistinguishable from a cell that has acquired the capability by a fixed and permanent mutation. Thus, in the case of a chronic infection, the cell behaves as if it has gained the capability, and this will facilitate the acquisition of other capabilities. This is the basis of the action of a tumour promoter, and this behaviour is mirrored very clearly by *Helicobacter*, which is now recognised as a carcinogen (Table 2).

While the generation of a cancer may require multiple competencies, it is not necessary for a putative carcinogenic factor to have the ability to influence all 8 capabilities for it to be classed as a carcinogenic agent. In this regard, it is evident that the action of PMT when tested against the Hanahan and Weinberg hallmarks shows evidence of being able to influence some, but not all capabilities.

It is very clear that PMT has the ability to mimic the first requirement, namely to produce sustained proliferative signalling. Indeed in experiments where cells were treated with PMT and then replated in the absence of toxin, enhanced growth continued for a number of days (Rozengurt et al. 1990), which presumably reflects the sustained effect of toxin that remained within confluent cells.

There is less hard molecular evidence to discuss the ability of PMT to facilitate the cell in evading growth suppressors. Nevertheless, it is known that PMT can overcome the normal check of contact inhibition and trigger confluent quiescent cells to undergo rounds of proliferation (Rozengurt et al. 1990), and to grow in an anchorage independent manner (Higgins et al. 1992). Similarly, the activation of β-catenin linked pathways would be predicted to activate the *c-myc* gene, and it has been shown that *c-myc* expression is enhanced by PMT treatment (Wilson et al. 2000).

There is evidence that PMT inhibits apoptosis (Preuß et al. 2010), but the concept that prolonged exposure to PMT might affect replicative immortality has not been tested. Likewise there is no evidence for or against the concept that PMT affects angiogenesis. PMT displays some properties that would facilitate invasion and metastasis, notably the ability to promote anchorage independent growth (Higgins et al. 1992).

Hanahan and Weinberg (2011) discuss two emerging hallmarks, namely deregulation of cellular energetics and immune avoidance. There have been no investigations whether PMT has any direct effect on metabolism, or to support a role in immune avoidance. Hildebrand et al. (2010) have demonstrated that the PMT-induced increase in STAT activity activates the Pim-1 serine/threonine kinase. The authors point out that this is the first example of a bacterial toxin activating this pathway, which has previously been shown to be a target for viral oncogenes and highlights a potentially noninflammatory oncogenic pathway. Hanahan and Weinberg (2011) also highlight a couple of enabling characteristics, namely genome instability and inflammation. There is no evidence for a direct mutational effect, and conflicting information on inflammation. The original publications on the effect of toxigenic infections or PMT treatment of animals provides no evidence of an inflammatory process (Foged et al. 1987; Hoskins et al. 1997), but the observation that the pro-inflammatory enzyme Cox-2 is activated (Orth et al. 2007) suggests the opposite.

It is therefore clear that PMT has the ability to lead to several of the Hanahan and Weinberg capabilities in a similar manner to the effects identified for *Helicobacter*.

4 A Role for PMT in Human Cancer?

The original discovery of PMT came from groups working on the porcine disease atrophic rhinitis (Pederson and Barfod 1981). A complication that delayed a full understanding of this disease was the observation that not all porcine strains of *P. multocida* expressed PMT (Foged et al. 1988). It is not known how widespread carriage of *P. multocida* is amongst pig populations. The disease atrophic rhinitis is not fatal, but is believed to lead to reduced growth.

There is very little data about *P. multocida* infections in humans. Most of the published information consists of individual case studies, none of which report on the toxigenic status of the isolated *P. multocida*. In most of these reports, it is thought that infection has arisen as a result of contact with companion animals, in particular cats. The most comprehensive study of human *P. multocida* infections analysed a bank of almost 1,000 isolates, but did not investigate their toxigenic status (Escande and Lion 1993). The bank of isolates has subsequently been destroyed. Only five research papers have assessed the toxigenic status of human isolates of *P. multocida*. Frandsen et al. (1990) identified toxigenic *P. multocida* in a variety of species, including one human isolate, but nothing further is known about the history of this isolate. A more comprehensive study examined 94 patient isolates from two French hospitals (Donnio et al. 1991). The majority of the isolates (53) came from pet-inflicted wounds and none of these was toxigenic. The remainder were respiratory isolates in patients associated with farming and six were toxigenic. One isolate was from a healthy carrier, four were associated with chronic bronchitis and one with infectious pneumopathy. A subsequent survey from the same team demonstrated that *P. multocida* carriage was more predominantly linked to pig breeders than cattle or rabbit breeders, but in this study none of the isolated strains was toxigenic (Donnio et al. 1994). It is of interest that an old report, years prior to the discovery of PMT, noted a linkage between chronic respiratory disease and the presence of *P. multocida* (Olsen and Needham 1952). A survey of clinical specimens, the majority from cat- or dog-inflicted wounds, identified only two toxigenic *P. multocida* isolates, one linked to a cat bite and one to a dog bite (Holst et al. 1992). A more recent survey of pigs and humans from the same geographical region found that around 10 % of the pig strains were toxigenic, while none of the human isolates was toxigenic (Marois et al. 2009).

The paucity of publications in this area precludes any firm conclusion being drawn about whether chronic carriage of toxigenic *P. multocida* occurs and whether it is linked to exposure to infected pigs. None of the studies looked for toxigenic *P. multocida* in both humans and the pigs that they were in contact with. However, given the known properties of PMT, it can certainly be hypothesised that chronic exposure to the toxin over a period of years is likely to predispose to cancer.

5 Conclusions

PMT has many properties that mark it out as having the potential to act as a tumour promoter. The G-proteins that it activates and many of their downstream signalling components are known to be upregulated or mutated in particular human cancers. While PMT does not directly initiate mutation, it has many similar properties to *H. pylori*, in terms of the signalling pathways that it impinges upon.

On the evidence so far, it would appear unlikely that PMT contributes significantly to human cancer. It is of interest that some human respiratory infections with *P. multocida* are suggested to be chronic, as this is a further characteristic that is shared by many carcinogenic agents. The PMT gene has been shown to reside on a prophage (Pullinger et al. 2004), which might enable it to be transmitted to other organisms, although there is currently no evidence that this occurs.

In conclusion, the PMT possesses many properties that suggest it could act as a potent tumour promoter, but it is unlikely to be a major cause of human pathology. Nevertheless, it represents an important new paradigm of how bacteria can promote cancer.

References

Agochiya M, Brunton VG, Owens DW, Parkinson EK, Paraskeva C, Keith WN, Frame MC (1999) Increased dosage and amplification of the focal adhesion kinase gene in human cancer cells. Oncogene 18:5646–5653

Aminova LR, Wilson BA (2007) Calcineurin-independent inhibition of 3T3-L1 adipogenesis by *Pasteurella multocida* toxin: suppression of Notch1, stabilization of β-catenin and pre-adipocyte factor 1. Cell Microbiol 9:2485–2496

Basu A, Sivaprasad U (2007) Protein kinase Cε makes the life and death decision. Cell Signal 19:1633–1642

Caygill CPJ, Braddick M, Hill MJ, Knowles RL, Sharp JCM (1995) The association between typhoid carriage, typhoid infection and subsequent cancer at a number of sites. Eur J Cancer Prev 4:187–193

Chikumi H, Vázquez-Prado J, Servitja J-M, Miyazaki H, Gutkind JS (2002) Potent activation of RhoA by $G\alpha_q$ and G_q-coupled receptors. J Biol Chem 277:27130–27134

Clark EA, Golub TR, Lander ES, Hynes RO (2000) Genomic analysis of metastasis reveals an essential role for RhoC. Nature 406:532–535

Da Silva L, Parry S, Reid L, Keith P, Waddell N, Kossai M, Clarke C, Lakhani SR, Simpson PT (2008) Aberrant expression of E-cadherin in lobular carcinomas of the breast. Am J Surg Pathol 32:773–783

Donnio PY, Avril JL, Andre PM, Vaucel J (1991) Dermonecrotic toxin production by strains of *Pasteurella multocida* isolated from man. J Med Microbiol 34:333–337

Donnio PY, Le Goff C, Avril JL, Pouedras P, Gras-Rouzet S (1994) *Pasteurella multocida*: oropharyngeal carriage and antibody response in breeders. Vet Res 25:8–15

Escande F, Lion C (1993) Epidemiology of human infections by *Pasteurella* and related groups in France. Zbl Bakt 279:131–139

Etienne-Manneville S, Hall A (2002) Rho GTPases in cell biology. Nature 420:629–635

Ferreri AJM, Dolcetti R, Magnino S, Doglioni C, Ponzoni P (2009) Chlamydial infection: the link with ocular adnexal lymphomas. Nat Rev Clin Oncol 6:658–669

Foged NT, Pedersen KB, Elling F (1987) Characterization and biological effects of the *Pasteurella multocida* toxin. FEMS Microbiol Lett 43:45–51

Foged NT, Nielsen JP, Pedersen KB (1988) Differentiation of toxigenic from nontoxigenic isolates of *Pasteurella multocida* by enzyme-linked immunosorbent assay. J Clin Microbiol 26:1419–1420

Foley JF, Singh SP, Cantu M, Chen L, Zhang HH, Farber JM (2010) Differentiation of human T cells alters their repertoire of G protein α-subunits. J Biol Chem 285:35537–35550

Franco AT, Israel DA, Washington MK, Krishna U, Fox JG, Rogers AB, Neish AS, Collier-Hyams L, Perez–Perez GI, Hatakeyama M, Whitehead R, Gaus K, O'Brien DP, Romero-Gallo J, Peek RM (2005) Activation of β-catenin by carcinogenic *Helicobacter pylori*. Proc Natl Acad Sci U S A 102:10646–10651

Frandsen PL, Foged NT, Petersen SK (1990) Characterisation of toxin from different Pasteurella multocida strains. In: Proceedings of International Pig Veterinary Society (11th Conference). Lausanne, Switzerland, p59

Golubovskaya VM (2010) Focal adhesion kinase as a cancer therapy target. Anti-Cancer Agent ME 10:735–741

Greenhough A, Smartt HJM, Moore AE, Roberts HR, Williams AC, Paraskeva C, Kaidi A (2009) The COX-2/PGE_2 pathway: key roles in the hallmarks of cancer and adaptation to the tumour microenvironment. Carcinogenesis 30:377–386

Hanahan D, Weinberg RA (2000) The hallmarks of cancer. Cell 100:57–70

Hanahan D, Weinberg RA (2011) Hallmarks of cancer: the next generation. Cell 144:646–674

Harmey D, Stenbeck G, Nobes CD, Lax AJ, Grigoriadis AE (2004) Regulation of osteoblast differentiation by *Pasteurella multocida* toxin (PMT): a role for Rho GTPase in bone formation. J Bone Miner Res 19:661–670

Hewavitharana T, Wedegaertner PB (2012) Non-canonical signaling and localizations of heterotrimeric G proteins. Cell Signal 24:25–34

Higgins TE, Murphy AC, Staddon JM, Lax AJ, Rozengurt E (1992) *Pasteurella multocida* toxin is a potent inducer of anchorage-independent cell growth. Proc Natl Acad Sci U S A 89:4240–4244

Hildebrand D, Walker P, Dalpke A, Heeg K, Kubatzky KF (2010) *Pasteurella multocida* toxin-induced Pim-1 expression disrupts suppressor of cytokine signalling (SOCS)-1 activity. Cell Microbiol 12:1732–1745

Holst E, Rollof J, Larsson L, Nielsen JP (1992) Characterization and distribution of *Pasteurella* species recovered from infected humans. J Clin Microbiol 30:2984–2987

Hoskins IC, Thomas LH, Lax AJ (1997) Nasal infection with *Pasteurella multocida* causes proliferation of bladder epithelium in gnotobiotic pigs. Vet Rec 140:22

Insel PA, Zhang L, Murray F, Yokouchi H, Zambon AC (2012) Cyclic AMP is both a pro-apoptotic and anti-apoptotic second messenger. Acta Physiol 204:277–287

Kalinec G, Nazarali AJ, Hermouet S, Xu N, Gutkind JS (1992) Mutated α subunit of the G_q protein induces malignant transformation in NIH 3T3 cells. Mol Cell Biol 12:4687–4693

Kleer CG, van Golen KL, Zhang Z-F, Rubin MA, Merajver SD (2002) Characterization of RhoC expression in benign and malignant breast disease. Am J Pathol 160:579–584

Krumins AM, Gilman AG (2006) Targeted knockdown of G protein subunits selectively prevents receptor-mediated modulation of effectors and reveals complex changes in non-targeted signaling proteins. J Biol Chem 281:10250–10262

Kuper H, Adami H-O, Trichopoulos D (2000) Infections as a major preventable cause of human cancer. J Inter Med 248:171–183

Lacerda HM, Lax AJ, Rozengurt E (1996) *Pasteurella multocida* toxin, a potent intracellularly acting mitogen, induces $p125^{FAK}$ and paxillin tyrosine phosphorylation, actin stress fiber formation, and focal contact assembly in Swiss 3T3 cells. J Biol Chem 271:439–445

Lagerström MC, Schiöth HB (2008) Structural diversity of G protein-coupled receptors and significance for drug discovery. Nat Rev Drug Discov 7:339–357

Lax AJ (2005) Bacterial toxins and cancer: a case to answer? Nat Rev Microbiol 3:343–349

Lax AJ, Chanter N (1990) Cloning of the toxin gene from *Pasteurella multocida* and its role in atrophic rhinitis. Microbiol 136:81–87

Lax AJ, Thomas W (2002) How bacteria could cause cancer: one step at a time. Trends Microbiol 10:293–299

Lee CH, Park D, Wu D, Rhee SG, Simon MI (1992) Members of the $G_q\alpha$ subunit gene family activate phospholipase $C\beta$ isozymes. J Biol Chem 267:16044–16047

Li G, Iyengar R (2002) Calpain as an effector of the Gq signaling pathway for inhibition of Wnt/β-catenin-regulated cell proliferation. Proc Natl Acad Sci U S A 99:13254–13259

Lyons J, Landis CA, Harsh G, Vallar L, Grünewald K, Feichtinger H, Duh Q-Y, Clark OH, Kawasaki E, Bourne HR, McCormick F (1990) Two G protein oncogenes in human endocrine tumors. Science 249:655–659

MacDonald BT, Tamai K, He X (2009) Wnt/β-catenin signaling: components, mechanisms, and diseases. Dev Cell 17:9–26

Marinissen MJ, Servitja JM, Offermanns S, Simon MI, Gutkind JS (2003) Thrombin protease-activated receptor-1 signals through G_q- and G_{13}-initiated MAPK cascades regulating c-Jun expression to induce cell transformation. J Biol Chem 278:46814–46825

Marois C, Fablet C, Gaillot O, Morvan H, Madec F, Kobisch M (2009) Molecular diversity of porcine and human isolates of *Pasteurella multocida*. J Appl Microbiol 107:1830–1836

Meigs TE, Fedor-Chaiken M, Kaplan DD, Brackenbury R, Casey PJ (2002) $G\alpha_{12}$ and $G\alpha_{13}$ negatively regulate the adhesive functions of cadherin. J Biol Chem 277:24594–24600

Meigs TE, Fields TA, McKee DD, Casey PJ (2001) Interaction of $G\alpha 12$ and $G\alpha 13$ with the cytoplasmic domain of cadherin provides a mechanism for β-catenin release. Proc Natl Acad Sci U S A 98:519–524

Mullan PB, Lax AJ (1996) *Pasteurella multocida* toxin is a mitogen for bone cells in primary culture. Infect Immun 64:959–965

Murphy AC, Rozengurt E (1992) *Pasteurella multocida* toxin selectively facilitates phosphatidylinositol 4,5-bisphosphate hydrolysis by bombesin, vasopressin, and endothelin. Requirement for a functional G protein. J Biol Chem 267:25296–25303

Narumiya S, Tanji M, Ishizaki T (2009) Rho signaling, ROCK and mDia1, in transformation, metastasis and invasion. Cancer Metastasis Rev 28:65–76

Naviglio S, Caraglia M, Abbruzzese A, Chiosi E, Di Gesto D, Marra M, Romano M, Sorrentino A, Sorvillo L, Spina A, Illiano G (2009) Protein kinase A as a biological target in cancer therapy. Expert Opin Ther Tar 13:83–92

Neves SR, Ram PT, Iyengar R (2002) G protein pathways. Science 296:1636–1639

Nomura A, Stemmermann GN, Chyou P-H, Kato I, Perez–Perez GI, Blaser MJ (1991) *Helicobacter pylori* infection and gastric carcinoma among Japanese Americans in Hawaii. New Engl J Med 325:1132–1136
Oldham WM, Hamm HE (2008) Heterotrimeric G protein activation by G-protein-coupled receptors. Nat Rev Mol Cell Biol 9:60–71
Olsen AM, Needham GM (1952) *Pasteurella multocida* in suppurative diseases of the respiratory tract. Am J Med Sci 224:77–81
Orth JHC, Lang S, Aktories K (2004) Action of *Pasteurella multocida* toxin depends on the helical domain of $G\alpha_q$. J Biol Chem 279:34150–34155
Orth JHC, Lang S, Taniguchi M, Aktories K (2005) *Pasteurella multocida* toxin-induced activation of RhoA is mediated via two families of Gα proteins, Gαq and Gα12/13. J Biol Chem 280:36701–36707
Orth JHC, Aktories K, Kubatzky KF (2007) Modulation of host cell gene expression through activation of STAT transcription factors by *Pasteurella multocida* toxin. J Biol Chem 282:3050–3057
Orth JHC, Fester I, Preuss I, Agnoletto L, Wilson BA, Aktories K (2008) Activation of $G\alpha_i$ and subsequent uncoupling of receptor-$G\alpha_i$ signaling by *Pasteurella multocida* toxin. J Biol Chem 283:23288–23294
Orth JHC, Preuss I, Fester I, Schlosser A, Wilson BA, Aktories K (2009) *Pasteurella multocida* toxin activation of heterotrimeric G proteins by deamidation. Proc Natl Acad Sci U S A 106:7179–7184
Oswald E, Nougayrède J-P, Taieb F, Sugai M (2005) Bacterial toxins that modulate host cell-cycle progression. Curr Opin Microbiol 8:83–91
Parsonnet J, Friedman GD, Vandersteen DP, Chang Y, Vogelman JH, Orentreich N, Sibley RK (1991) *Helicobacter pylori* infection and the risk of gastric carcinoma. New Engl J Med 325:1127–1131
Pederson KB, Barfod K (1981) The etiological significance of *Bordetella bronchiseptica* and *Pasteurella multocida* in atrophic rhinitis of swine. Nordisk Veterinaer Med 33:513–522
Preuß I, Hildebrand D, Orth JHC, Aktories K, Kubatzky KF (2010) *Pasteurella multocida* toxin is a potent activator of anti-apoptotic signalling pathways. Cell Microbiol 12:1174–1185
Pullinger GD, Bevir T, Lax AJ (2004) The *Pasteurella multocida* toxin is encoded within a lysogenic bacteriophage. Mol Microbiol 51:255–269
Radhika V, Dhanasekaran N (2001) Transforming G proteins. Oncogene 20:1607–1614
Roderick HL, Cook SJ (2008) Ca^{2+} signalling checkpoints in cancer: remodelling Ca^{2+} for cancer cell proliferation and survival. Nat Rev Cancer 8:361–375
Rozengurt E, Higgins T, Chanter N, Lax AJ, Staddon JM (1990) *Pasteurella multocida* toxin: potent mitogen for cultured fibroblasts. Proc Natl Acad Sci U S A 87:123–127
Russell W (1890) An address on a characteristic organism of cancer. Brit Med J 2.1356–1360
Rutter JM, Luther PD (1984) Cell culture assay for toxigenic *Pasteurella multocida* from atrophic rhinitis of pigs. Vet Rec 114:393–396
Rutter JM, Mackenzie A (1984) Pathogenesis of atrophic rhinitis in pigs: a new perspective. Vet Rec 114:89–90
Sabri A, Wilson BA, Steinberg SF (2002) Dual actions of the $G\alpha_q$ agonist *Pasteurella multocida* toxin to promote cardiomyocyte hypertrophy and enhance apoptosis susceptibility. Circ Res 90:850–857
Saha B, Kaur P, Tsao-Wei D, Naritoku WY, Groshen S, Datar RH, Jones LW, Imam SA (2008) Unmethylated E-Cadherin gene expression is significantly associated with metastatic human prostate cancer cells in bone. Prostate 68:1681–1688
Shah MH, Sainger RN, Telang SD, Pancholi GH, Shukla SN, Patel PS (2009) E-Cadherin truncation and Sialyl Lewis-X overexpression in oral squamous cell carcinoma and oral precancerous conditions. Neoplasma 56:40–47
Sparks AB, Morin PJ, Vogelstein B, Kinzler KW (1998) Mutational analysis of the APC/β-catenin/Tcf pathway in colorectal cancer. Cancer Res 58:1130–1134
Stabel S, Parker PJ (1991) Protein kinase C. Pharmacol Ther 51:71–95

Staddon JM, Chanter N, Lax AJ, Higgins TE, Rozengurt E (1990) *Pasteurella multocida* toxin, a potent mitogen, stimulates protein kinase C-dependent and -independent protein phosphorylation in Swiss 3T3 cells. J Biol Chem 265:11841–11848

Staddon JM, Barker CJ, Murphy AC, Chanter N, Lax AJ, Michell RH, Rozengurt E (1991) *Pasteurella multocida* toxin, a potent mitogen, increases inositol 1,4,5-trisphosphate and mobilizes Ca^{2+} in Swiss 3T3 cells. J Biol Chem 266:4840–4847

Switzer WP, Farrington DO (1975) Infectious atrophic rhinitis. In: Dunne HW and Leman AD (eds) Diseases of swine, 4th edn. Iowa State University Press, Ames, pp. 687–711

Tanabe S, Kreutz B, Suzuki N, Kozasa T (2004) Regulation of RGS-RhoGEFs by Gα12 and Gα13 proteins. Methods Enzymol 390:285–294

Tegtmeyer N, Wessler S, Backert S (2011) Role of the cag-pathogenicity island encoded type IV secretion system in *Helicobacter pylori* pathogenesis. FEBS J 278:1190–1202

Thomas W, Pullinger GD, Lax AJ, Rozengurt E (2001) *Escherichia coli* cytotoxic necrotizing factor and *Pasteurella multocida* toxin induce focal adhesion kinase autophosphorylation and src association. Infect Immun 69:5931–5935

Titus B, Schwartz MA, Theodorescu D (2005) Rho proteins in cell migration and metastasis. Crit Rev Eukar Gene 15:103–114

Van Raamsdonk CD, Bezrookove V, Green G, Bauer J, Gaugler L, O'Brien JM, Simpson EM, Barsh GS, Bastian BC (2009) Frequent somatic mutations of GNAQ in uveal melanoma and blue naevi. Nature 457:599–603

Voyno-Yasenetskaya TA, Pace AM, Bourne HR (1994) Mutant α subunits of G12 and G13 proteins induce neoplastic transformation of Rat-1 fibroblasts. Oncogene 9:2559–2565

Ward PN, Miles AJ, Sumner IG, Thomas LH, Lax AJ (1998) Activity of the mitogenic *Pasteurella multocida* toxin requires an essential C-terminal residue. Infect Immun 66:5636–5642

Welton JC, Marr JS, Friedman SM (1979) Association between hepatobiliary cancer and typhoid carrier state. Lancet 313:791–794

White BD, Chien AJ, Dawson DW (2012) Dysregulation of Wnt/β-catenin signaling in gastrointestinal cancers. Gastroenterology 142:219–232

Wilson BA, Zhu X, Ho M, Lu L (1997) *Pasteurella multocida* toxin activates the inositol triphosphate signaling pathway in *Xenopus* oocytes via $G_q\alpha$-coupled phospholipase C-β1. J Biol Chem 272:1268–1275

Wilson BA, Aminova LR, Ponferrada VG, Ho M (2000) Differential modulation and subsequent blockade of mitogenic signaling and cell cycle progression by *Pasteurella multocida* toxin. Infect Immun 68:4531–4538

Woll PJ, Rozengurt E (1988) Bombesin and bombesin antagonists: studies in Swiss 3T3 cells and human small cell lung cancer. Brit J Cancer 57:579–586

Wotherspoon AC, Ortiz-Hidalgo C, Falzon MR, Isaacson PG (1991) *Helicobacter pylori*-associated gastritis and primary B-cell gastric lymphoma. Lancet 338:1175–1176

Wu S, Morin PJ, Maouyo D, Sears CL (2003) *Bacteroides fragilis* enterotoxin induces c-myc expression and cellular proliferation. Gastroenterology 124:392–400

Yanamadala V, Negoro H, Denker BM (2009) Heterotrimeric G proteins and apoptosis: intersecting signaling pathways leading to context dependent phenotypes. Curr Mol Med 9:527–545

Zhao Y, He D, Stern R, Usatyuk PV, Spannhake EW, Salgia R, Natarajan V (2007) Lysophosphatidic acid modulates c-Met redistribution and hepatocyte growth factor/c-Met signaling in human bronchial epithelial cells through PKC δ and E-cadherin. Cell Signal 19:2329–2338

Zywietz A, Gohla A, Schmelz M, Schultz G, Offermanns S (2001) Pleiotropic effects of *Pasteurella multocida* toxin are mediated by G_q-dependent and -independent mechanisms - involvement of G_q but not G_{11}. J Biol Chem 276:3840–3845

Index

Current Topics in Microbiology and Immunology (2012) 361: 145–147
DOI: 10.1007/978-3-642-31017-1

MIX
Papier aus verantwortungsvollen Quellen
Paper from responsible sources
FSC® C105338

If you have any concerns about our products,
you can contact us on
ProductSafety@springernature.com

In case Publisher is established outside the EU,
the EU authorized representative is:
Springer Nature Customer Service Center GmbH
Europaplatz 3, 69115 Heidelberg, Germany

Printed by Libri Plureos GmbH
in Hamburg, Germany